Artificial Intelligence and Machine Learning Foundations is a thought-provoking exploration of AI's evolution, its core workings and practical applications. The case studies and toolbox are invaluable resources for anyone in the field.
Darren Winter, *Director, Duco Digital Training and PhD Student in AI Ethics*

This book is a highly recommended read for those starting out on Artificial Intelligence learning or as a refresher, written by two very knowledgeable authors, who put across the subject in an easy-to-understand manner.
Mat Gardam, *Cyber Security Professional, Armoured-Cyber Ltd*

This is a great book for executives to understand how AI may fit into their company and the types of benefits that they may gain and some of the pitfalls they may fall into. The book gives a clear picture of where AI fits with other types of decision-making processes. It covers many different aspects of how AI can be developed and how it affects society, with a particular focus on ethics and the human machine interface.
Dr Paul Edward Mort, *B.Eng. (Hons), PhD, MBA, MIMechE CEng FNucl, Lead Consultant for PM Advanced Technologies Consultancy Ltd*

Many commentators see AI as the key enabler of the fifth industrial revolution. In this clear and concise guide to the BCS Artificial Intelligence Foundation qualification, Andrew Lowe and Steve Lawless set out to clearly explain the core concepts around artificial intelligence, machine learning, applications in robotics and more in this new edition. A 'must have' guide in preparation for the newly updated BCS AI Foundation exam, this book will not only assist with exam success but will capture your imagination to the possibilities in front of us when reimagining the human vs machine relationship.
Richard Webber, *Digital Transformation Manager*

Reviews on previous edition:

This book is a welcome read for those who want to know more about AI but don't have a degree level background in the underlying subjects. It successfully brings the beginner up to speed and is written with such enthusiasm that the more complex topics become manageable. For a foundation level book it covers an impressive content range, whilst also providing a solid grounding in machine learning. The topics are clearly and succinctly exemplified by three case studies which bring AI into the real world.
Rosie Sheldon, *Senior Trainer, TSG-training*

Produces a great storm of ideas and thoughts with a funny touch from the authors. This is a piece of art and a 'must read' for good immersion in the theme.
David Mondragón Tapia, *IT & Business Consultant, DieresiS – Business and Professional Services*

Manages to provide a fundamental grounding in the theory and application of AI without relying on mathematical or data science skills ... the theory is colourfully illustrated with examples of both AI successes and challenges, which will prove an invaluable foundation to the reader's learning from experience in the world of artificial intelligence.
Mark Ainsworth, *Director and Business Analyst, Promising ICT Limited*

ARTIFICIAL INTELLIGENCE AND MACHINE LEARNING FOUNDATIONS

BCS, THE CHARTERED INSTITUTE FOR IT

BCS, The Chartered Institute for IT, is committed to making IT good for society. We use the power of our network to bring about positive, tangible change. We champion the global IT profession and the interests of individuals, engaged in that profession, for the benefit of all.

Exchanging IT expertise and knowledge
The Institute fosters links between experts from industry, academia and business to promote new thinking, education and knowledge sharing.

Supporting practitioners
Through continuing professional development and a series of respected IT qualifications, the Institute seeks to promote professional practice tuned to the demands of business. It provides practical support and information services to its members and volunteer communities around the world.

Setting standards and frameworks
The Institute collaborates with government, industry and relevant bodies to establish good working practices, codes of conduct, skills frameworks and common standards. It also offers a range of consultancy services to employers to help them adopt best practice.

Become a member
Over 70,000 people including students, teachers, professionals and practitioners enjoy the benefits of BCS membership. These include access to an international community, invitations to a roster of local and national events, career development tools and a quarterly thought-leadership magazine. Visit www.bcs.org to find out more.

Further information
BCS, The Chartered Institute for IT,
3 Newbridge Square,
Swindon, SN1 1BY, United Kingdom.
T +44 (0) 1793 417 417
(Monday to Friday, 09:00 to 17:00 UK time)
www.bcs.org/contact

http://shop.bcs.org/
publishing@bcs.uk

www.bcs.org/qualifications-and-certifications/certifications-for-professionals/

ARTIFICIAL INTELLIGENCE AND MACHINE LEARNING FOUNDATIONS

Learning from experience
Second edition

Steve Lawless and Andrew Lowe

Published by BCS Learning and Development Ltd, a wholly owned subsidiary of BCS, The Chartered Institute for IT, 3 Newbridge Square, Swindon, SN1 1BY, UK.
www.bcs.org

Paperback ISBN: 978-1-78017-6734
PDF ISBN: 978-1-78017-6741
ePUB ISBN: 978-1-78017-6758

Ebook available

British Cataloguing in Publication Data.
A CIP catalogue record for this book is available at the British Library.

Disclaimer:
The views expressed in this book are of the authors and do not necessarily reflect the views of the Institute or BCS Learning and Development Ltd except where explicitly stated as such. Although every care has been taken by the authors and BCS Learning and Development Ltd in the preparation of the publication, no warranty is given by the authors or BCS Learning and Development Ltd as publisher as to the accuracy or completeness of the information contained within it and neither the authors nor BCS Learning and Development Ltd shall be responsible or liable for any loss or damage whatsoever arising by virtue of such information or any instructions or advice contained within this publication or by any of the aforementioned.

All URLs were correct at the time of publication.

Publisher's acknowledgements
Reviewers: Harriet Rogers, Nicolas Sinnott and Jude Umeh
Publisher: Ian Borthwick
Commissioning editor: Heather Wood
Production manager: Florence Leroy
Project manager: Sunrise Setting Ltd
Copy-editor: Mary Hobbins
Proofreader: Barbara Eastman
Indexer: David Gaskell
Cover design: Alex Wright
Cover image: iStock/metamorworks
Sales director: Charles Rumball
Typeset by Lapiz Digital Services, Chennai, India

CONTENTS

FIGURES AND TABLES

AUTHORS

Steve Lawless – Steve has always had a keen interest in science and technology and has worked in the computer industry for over 40 years. Steve has trained literally thousands of people to give them the skills they need to make computers work. He knows Andy from volunteering at the local ski club. He runs a successful training company that has a worldwide customer base. He's written over 100 training courses on IT and enjoys making technology easy to understand and accessible.

Dr Andrew Lowe – Andy is an engineer specialising in using computers to solve the challenges engineers face. Learning from experience and learning from data have been a fundamental part of his career and this helps him in his day-to-day work. Some of the problems he's worked on can only be solved with supercomputers. He obtained his PhD from Cambridge and has worked in academia and industry. He's also helped young AI start-up companies gain traction.

ACKNOWLEDGEMENTS

We would like to thank our families who have been patient. It's amazing how much time it takes to pull together learning from experience in an easy to understand book. The technical side of AI can be challenging, so the numerous teachers, lecturers, supervisors, tutors and mentors are too many to mention individually. Thank you, you know who you are, especially those who have made their lectures and notes available on the web so everyone can learn.

Dr Paul Mort who worked at Sellafield Sites Ltd is a keen enthusiast and supporter of the AI Essentials and AI Foundation courses. His challenging questions while reviewing the courses made us think very carefully and we thank him. Paul is a roboticist, and we were highly aware that AI is much more enriching and a broader subject than machine learning. Indeed, we left the digital revolution behind in the third industrial revolution and we are now well on our way to the fifth. The AI agent is a really important and fundamental concept that we can easily overlook if we think AI is a digital machine learning technology.

BCS have played a big role in bringing this book and the courses together. We'd like to thank, in no particular order, Ian Borthwick, Heather Wood, Tope Gay, Gillian Cullen and James Benefield. There have been many others involved we are sure. Thank you.

Wikipedia has also played a big role; wherever possible we've made access to information and references easy. Thank you to all those authors who have put their work on the web for everyone to learn from.

We hope you find something useful in this book and, above all, are sceptical, critical and rigorous in your study, but we want you to learn from experience.

We are human – we can make mistakes and would love to hear all feedback good and bad. This is part of the scientific method and learning from experience. Contact us via LinkedIn: www.linkedin.com/in/stevelawless.

ABBREVIATIONS

3D	three-dimensional
ABM	agent-based modelling
AGI	artificial general intelligence
AGV	automated guided vehicle
AI	artificial intelligence
AIMS	artificial intelligence management system
AIOps	artificial intelligence to IT operations
ANN	artificial neural network
AR	augmented reality
ASIMO	Advanced Step in Innovative Mobility
CNN	convolutional neural network
cobots	collaborative robots
DARPA	Defense Advanced Research Projects Agency (USA)
DBN	deep belief network
DNN	deep neural network
DPA	Data Protection Act
EQ	emotional quotient
EU	European Union
FAT/ML	fairness, accountability and transparency in ML
GAN	generative adversarial network
GDPR	General Data Protection Regulation
Gen AI	generative AI
GOMS	goals, operators, methods and selection rules
GPAI	general purpose artificial intelligence
GPT	generative pre-trained transformer
GPU	graphics processing unit
HCAI	human-centred artificial intelligence
HCI	human–computer interaction
HITL	human in the loop
IA	intelligent automation or augmentation

ICT	information and communications technology
IEC	International Electrotechnical Commission
IoT	Internet of Things
ISO	International Organization for Standardization
KNN	K-nearest neighbour
LSTM	long short-term memory
MHP	model human processor
ML	machine learning
MLOps	machine learning operations
MLP NN	multilayer perceptron neural network
ModelOps	model operations
NI	natural intelligence
NIST	National Institute of Standards
NLP	natural language processing
NMT	neural machine translation
NN	neural network
OCR	optical character recognition
PDF	probability density function
PESTLE	political, economic, sociological, technological, legal and environmental
RBM	restricted Boltzmann machine
RNN	recurrent neural network
ROI	return on investment
RPA	robotic process automation
SVM	support vector machine
SWOT	strengths, weaknesses, opportunities, threats
TAM	technology acceptance model
UCD	user-centred design
UX	user experience
VAE	variational autoencoder
VR	virtual reality
WCAG	Web Content Accessibility Guidelines
XAI	explainable AI

PREFACE

This book was written with the express purpose of supporting the BCS Essentials Certificate in Artificial Intelligence, the BCS Foundation Certificate in Artificial Intelligence training courses and other scheduled courses already under development in the BCS AI course pipeline.

Its aim is to document what artificial intelligence and machine learning are and what they are not, separate fact from fiction and educate those with an interest in AI and machine learning. We have included a number of topics that introduce the basics of machine learning, and ethics.

We believe that this book is relatively unique in that it brings together information and concepts in one book that are, until now, spread across numerous other volumes. The book also aims to simplify (where possible) complex and confusing concepts, making the topic highly accessible to those without a high-level degree in the subjects covered.

Stuart Russell and Peter Norvig, the authors of the standard *Artificial Intelligence* textbook (Russell and Norvig, 2016) explain that AI is a universal subject and helpful to us all. Learning from experience is AI's signature. We will use this a lot, it applies to machines as it does, perhaps more so, to humans. Einstein is often quoted as saying, 'The only source of knowledge is the experience'.

Ask yourself the question, where's the learning from experience and can machines help? If the answer to this question is yes, then AI can help you!

Our hope here is to bring these concepts to life by balancing theory with practice. We want to make the human part of an AI project as important as the AI itself. After all, machines are here to take the heavy lifting away from us humans. Not only that, but to give us extra capabilities that we could not do by ourselves. Humans and machines have unique capabilities and it is important to find the right balance. We have not been tempted to write a machine learning course which is very popular at the moment. If we did, we would somehow have to stretch, contort and manipulate it to explain concepts like consciousness and robotics. Even explaining narrow and strong artificial general intelligence (AGI) would be tricky. And to embrace the practical human side would be virtually impossible. BCS is also a chartered institute that can certify engineers to be chartered and hold the title of CEng. This course goes right back to fundamentals so engineers, scientists and mathematicians can gain some traction in AI.

Scientists, mathematicians and engineers all use machine learning but probably don't call it this. All of these professions operate ethically and so should AI. People, society and governments are quite rightly concerned about AI and its potential. As such, we have adopted the EU guidelines on the ethical use of AI. In fact, the EU guidelines give us two practical benefits:

- to measure the success of our AI projects by concentrating on Design for All and sustainability;
- to use the metrics from the first part to build reasoning that can be used in planning and business cases.

The EU guidelines for AI ask us to build a human-centric ethical purpose that gives us trustworthy and technically robust AI. Stuart Russell's recent book on human compatible AI (2019) puts the human into our consideration when undertaking an AI project. His take on this asks if we should think about AI as serving our needs, telling us how to be better humans. In doing so, it gives us an alternative to controlling AI by asking the AI what is best for us as humans. In doing so, this elegant idea aims to solve the control problem of AI.

As we move into the fifth industrial revolution, we have the opportunity to think about humans and machines. How do we complement each other? How can AI and machines leave humans to undertake more valued work and deal with ambiguity or contradictions? To become more human, build better societies and for all to exploit their talents. What are the new roles for humans, humans and machines and machines only? Simply thinking of these roles, focusing on opportunities, leaves a richer environment as we progress, less distracted by robots coming to take over our jobs! Can you, your current career or organisation benefit from learning from experience? If so, read on.

1 FOUNDATIONS OF ARTIFICIAL INTELLIGENCE

1.1 UNDERSTANDING ARTIFICIAL INTELLIGENCE

In this chapter we are going to make sense of artificial intelligence (AI). In particular, framing machine learning (ML) and AI into the concept of a learning agent that maps onto a simple description of human intelligence. AI is about creating intelligent entities (Russell and Norvig, 2016) and ML is about learning from data (Mitchell, 2018).

1.1.1 Definitions of key AI terms

1.1.1.1 Human intelligence
Human intelligence is the culmination of billions of years of evolution from single cell organisms to what we are today, which is ultimately marked by our ability to undertake complex mental feats and be self-aware. It also includes the ability to recognise our place in the universe and ask annoying philosophical questions such as 'Why are we here?' and 'What is our purpose?'

There are many definitions of human intelligence. Our chosen definition is useful because it is intuitive and gives us a practical base that builds a strong foundation for AI. In fact, it needs to be a little more than intuitive: it also needs to guide us as to what AI is useful for in practice. When considering this definition, we must keep in the back of our minds that the need to find the right balance of theory and practice is paramount. We will also need to understand that AI and ML have significant limitations, and this will become apparent as we move through the book.

Take five minutes to think about what it means to be human.

Leonardo da Vinci captured the essence of his science in art (Wikipedia Contributors, 2024c). René Descartes stated *'cogito, ergo sum'* (I think, therefore I am) (Wikipedia Contributors, 2024b). Ada Lovelace wrote the first algorithm and notes on the role of humans and society with technology (Wikipedia Contributors, 2024a). Neil Armstrong was the first to set foot on the moon, and changed our perspective of the world completely (Wikipedia Contributors, 2024d). Roger Bannister was the first to run a mile in under four minutes (Wikipedia Contributors, 2024h). Dr Karen Spärck Jones gave us the theoretical foundations of the search engine (Wikipedia Contributors, 2024e). Tu Youyou is a tenacious scientist who discovered a cure for malaria, which won her the Nobel Peace Prize in 2015 (Wikipedia Contributors, 2024f). Tu's intellectual talents are amazing, and, after perfecting her cure, she volunteered to be the first person for it to be tested on. We could ask ourselves if this was confidence or bravery.

We have set ourselves up here to introduce the concept of subjectivity. We have free will and all of us have our own unique subjective experience. We are conscious and subjective, and conscious experience is something that we will need to be all too aware of as we develop AI.

Robert Sternburg gives us a useful definition of human intelligence (Sternburg, 2017), at least in so far as it relates to AI:

> Human intelligence: mental quality that consists of the abilities to learn from experience, adapt to new situations, understand and handle abstract concepts, and use knowledge to manipulate one's environment.

Here we can quickly recognise the desire to manipulate our environment in some way, and that we will use all our human talents to do so. This is a general desire so we still need to identify the type of learning from experience that humans do, or, to put it another way, what machines can help us with.

Sometimes called natural intelligence (NI), human intelligence is generally considered to be the intellectual accomplishment of humans and has been discussed by philosophers for thousands of years. Of course, other living things possess NI to some degree as well, but for now let's just consider human intelligence. We are biased, but it's fair to say that humans are the most intelligent living organisms on the planet. Just as we developed tools in the Stone Age and Iron Age, we are now equipping ourselves with machines to help us intellectually.

We may phrase this as coming to the correct conclusions – hypothesising and testing – and understanding what is real, although we sometimes get it wrong. It's also about how to understand complex problems such as weather prediction or winning a game of chess; adapting what we have learned through things like abstraction, induction, simplification and creativity. It allows us to adapt and control our environment and interact socially, giving us an evolutionary advantage.

It may also make sense to consider human intelligence from a number of perspectives, such as:

- Linguistic intelligence – the ability to communicate complex ideas to another.
- Mathematical intelligence – the ability to solve complex problems.
- Interpersonal intelligence – the ability to see things from the perspective of others, or to understand people in the sense of having empathy.

So, how do we acquire these particular skills or traits? Through learning from experience. Human learning is the process of acquiring knowledge. It starts at a very early age, perhaps even before we are born. Our behaviour, skills, values and ethics are acquired and developed when we process information through our minds and learn from those experiences. Human learning may occur as part of education, personal development or any other informal/formal training, and is an ongoing process throughout our lives. Each person has a preference for different learning styles and techniques (e.g. visual, aural, kinaesthetic, etc.).

1.1.1.2 The scientific method

Since the 1980s, AI has adopted the scientific method (Russell and Norvig, 2016). The scientific method (Wikipedia Contributors, 2024g) is a systematic, empirical approach used for acquiring new knowledge or correcting and integrating previous knowledge. It relies on empirical evidence and the testing of hypotheses through controlled experiments and observations. This method is fundamental to scientific inquiry and is structured to minimise biases and maximise objectivity and reproducibility. Wikipedia makes this more explicit by identifying the following:

- careful observation;
- rigorous scepticism;
- hypothesis formulation;
- experiments for testing;
- hypothesis refinement.

The process begins with careful observation. Scientists observe phenomena and gather data, noting patterns and irregularities. These observations often lead to questions (rigorous scepticism) about how or why something occurs. For instance, a biologist might notice a particular plant species thriving in one environment but not another and wonder what factors contribute to its success.

Following observation, scientists formulate a hypothesis. A hypothesis is a testable statement that proposes a potential explanation for the observed phenomenon. It should be specific and measurable. For example, the biologist might hypothesise that the plant thrives due to higher nitrogen levels in the soil.

The next step is experimentation. Experiments are designed to test the hypothesis under controlled conditions. They should include a control group that does not receive the experimental treatment and one or more experimental groups that do. The goal is to isolate the variable being tested to determine its effect. Continuing with the example, the biologist could plant the species in soils with varying nitrogen levels while keeping other factors constant.

After conducting experiments, scientists engage in data analysis. They collect and examine the data to see if they support or refute the hypothesis. Statistical methods are often employed to determine the significance of the results.

Based on the data analysis, scientists reach a conclusion. If the hypothesis is supported, it can lead to further experimentation and refinement. If it is not supported, the hypothesis may be rejected or modified.

An important aspect of the scientific method is replication. Results must be reproducible by other researchers following the same methodology. This ensures that findings are not due to chance or specific experimental conditions. Peer review and publication in scientific journals facilitate this by allowing other scientists to scrutinise and replicate the work.

Finally, the scientific method is an iterative process. New questions often arise from the conclusions of one study, leading to new hypotheses and experiments. This cycle of inquiry builds a more robust understanding of the natural world. Over time, hypotheses that consistently withstand rigorous testing may contribute to the development of scientific theories, which are comprehensive explanations of natural phenomena based on extensive evidence.

The scientific method, therefore, is not just a series of steps but a dynamic and ongoing process of inquiry that drives scientific progress. It underscores the importance of scepticism, critical thinking and the constant quest for deeper understanding. Through this method, science continuously evolves, expanding our knowledge and capability to address complex questions and challenges. Later, we will explore how this has led to the industrial revolutions – we are approaching the fifth at a great pace.

1.1.1.3 Artificial intelligence
In simple terms, AI is intelligence demonstrated by machines, in contrast to the NI displayed by humans and other animals. AI is about creating intelligent entities.

Stuart Russell and Peter Norvig, the authors of the standard AI textbook *Artificial Intelligence: A Modern Approach* (2016), explain that AI is a universal subject and helpful to us all. Learning from experience is AI's signature. We will use this concept a lot; it applies to machines as much as it does to humans, perhaps more so.

Einstein is often quoted as saying: 'The only source of knowledge is the experience.' Ask yourself: can machines help us to learn from experience? If the answer to this question is yes, then AI can help you.

1.1.1.4 Machine learning
In this section, we need to make the distinction between ML and AI. This is no easy task until we remind ourselves that AI is a universal subject that can help any pursuit of learning from experience to achieve a goal (Russell and Norvig, 2016). ML is only part of this – the distinction starts with the definitions of an AI agent and ML. Tom Mitchell's ML definition is the one usually quoted (Mitchell, 2018); it relates well to digital computing that is part of all our lives.

> A computer program is said to learn from experience, *E*, with respect to some class of tasks, *T*, and performance measure, *P*, if its performance at tasks in *T*, as measured by *P*, improves with experience, *E*.

AI is about intelligent entities, or AI agents, interacting with an environment to achieve a goal. Not just one entity but multiple. AI is about humans and machines, how they interact, how they learn, what they experience. AI asks us to be explicit about humans achieving their goals. Can we build machines that we can work with to improve us as humans? AI asks hard questions of consciousness, philosophy, ethics and science. It deals with complex phenomena that we do not currently have the machines to explore. AI of the future is about how humans and machines will co-exist. In 2019 Stuart Russell released his latest book on human-compatible AI, *Human Compatible: Artificial Intelligence and the Problem of Control* (2019), which captures the true nature of AI and how, used wisely, we can benefit from a future of humans and machines.

As we are coming to realise the benefit of digital computing ML, this will unlock the potential of AI in other areas, such as:

- engineering and building intelligent entities;
- medicine and improving health- and social-care;
- business analytics, and others.

When we simply think about products, representing the world in a digital simulation – or what is (at the moment) ones and zeros – is not ideal. Digital computation has its limitations, and we need better machines. We have run out of digital processing power. AI is much more than high-performance computing and programming of machines that deal with ones and zeros – digitally simulated pizza doesn't taste of anything, and digital weather does not get us wet.

The AI machines of the future will incorporate digital computers, but, when we think about it, it's actually hard to represent the mathematical operations we need. We are limited by the processing power, energy and accuracy of today's technology – and, as a result, we can only concentrate on narrow ML, focused on specific, well-defined tasks or goals.

The examples given of these types of tasks are playing games such as chess, checkers and draughts. Modern-day games include simulation games and very advanced strategic, well-engineered games like Go; these types of games can be explicitly defined on a digital computer. Practical examples in the real world include optimising where aircraft park at an airport or the logistics of delivering a parcel; again, a reasonably well-ordered and engineered environment in which ML can optimise something.

ML is focused on explicitly defining a problem that can be solved on a computer. These problems can be complicated, non-linear and statistical. If we are to use ML in our AI, we must be able to represent our problem mathematically and in such a way that it can be solved by a machine. Today, we typically use digital computers. However, quantum, analogue, optical and biological computers are on their way.

THE REAL WORLD ISN'T A DIGITAL SIMULATION

What would happen if we designed everything digitally? Would we trust something that is designed entirely by digital modelling? In this box we are not going to concentrate on a single example, but look at a concept that has been around for about 20 years – that of the digital twin.

The digital twin is an example of where AI can be used to learn from a digital simulation, and maybe sensors; there are many examples where this concept has been used successfully (Wikipedia Contributors, 2024i). We also need to note that just because we have simulated something, it doesn't mean it will perform the same way in reality.

The digital twin has various definitions, and is not something to start without considering carefully if it is a good representation of the real world. Even a simple model of the Earth's weather turns out to be incredibly complicated and requires large supercomputers to undertake.

AI can also draw on the system of systems approach to modelling complex systems – lots of subsystems that, when combined, develop ever more complex behaviour. This has the advantage of being an academic subject linked closely with the fundamental disciplines that underpin AI and ML. Examples of these approaches are agent-based modelling and lifecycle analysis used in sustainability. With a system of systems approach, you can break the larger problem down into manageable chunks. Indeed, we might find out that one system requires so much computing power that the larger digital twin might not be feasible, or indeed realistic. In this case we need to build a model of the subsystem that can be incorporated into the system of systems approach.

The digital twin concept built using the system of systems approach might be useful in ensuring that our AI is human-centric, technically robust and trustworthy.

Digital ML has become very popular recently with the success of convolutional deep neural networks (NNs) and, more recently, generative AI. These numerical techniques have given us an understanding of how the human mind solves problems. Just to clarify this, the operation of the human eye led to the breakthrough in NNs: the convolution neural network (CNN). So, when we think about ML, we can think about an AI agent learning from data. These machines are now so good at playing games that they can beat the world champion at these types of games; we look at this in more detail in Section 1.1.2.1.

The AI agent is much more than a narrowly focused computer program. ML works on data in the computer. We must work really hard to think of this as an interface with actuators and sensors in an environment. It is sensible to think, here, that ML learns from data in a computational environment. This is a good starting point to opening up the world of AI. AI is about humans and machines working together to achieve goals. We might even go on to say that ML is an AI enabler setting the foundation for a future of humans and machines.

1.1.1.5 Deep learning and generative AI
Deep learning and generative AI represent significant milestones in the journey of narrow ML; a step change in how we interact with technology to create new realms of possibility in art and science – these technologies are reshaping our world. As research progresses, we can anticipate even more innovative applications and breakthroughs, making deep learning and generative AI integral components of our future.

Deep learning has the ability to model and understand complex patterns in data. They are built on the use of artificial NNs that date back nearly a hundred years. In essence, they are a simple mathematical model of the human brain. Deep learning models consist of multiple layers, including an input layer, several hidden layers and an output layer. Each layer contains nodes (neurons) that process and transform data. The depth of the network (i.e. the number of layers) enables it to learn hierarchical representations of data, capturing intricate features at different levels of abstraction. The term 'deep' captures the large number of layers we see in the large model.

Notable applications include autonomous vehicles, where deep learning algorithms enable perception and decision-making, and healthcare, where deep learning assists in diagnosing diseases and predicting patient outcomes. Moreover, deep learning models power recommendation systems, enhancing user experiences on platforms such as Netflix and Amazon. In natural language processing (NLP), recurrent neural networks (RNNs) and transformers have significantly improved machine translation, sentiment analysis and language generation.

Generative AI creates new content rather than analysing existing data. This is achieved through generative models, which learn the underlying distribution of data and can produce novel samples that resemble the training data. Two prominent types of generative models are generative adversarial networks (GANs) and variational autoencoders (VAEs). GANs have been instrumental in generating high-quality images, videos and even music. They have applications in art, fashion and content creation, allowing for the synthesis of new, creative works. Here we see the immediate challenge of the ethical and legal issues this causes.

AI foundation models are large-scale ML models trained on vast amounts of diverse data to perform a wide array of tasks. They have been developed within the AI industry, with academia now doing the rigorous fundamentals (Bommasani et al., 2021). These models, such as GPT-4, are designed to understand and generate human-like text by leveraging deep learning techniques and extensive NN architectures. They serve as a foundational base for various AI applications, including NLP, image recognition and decision-making systems. By being pre-trained on extensive data sets, foundation models can be fine-tuned for specific applications with relatively smaller amounts of data, making them highly adaptable and efficient. Their broad capabilities and scalability have unlocked the field of AI, enabling more sophisticated and versatile AI solutions.

1.1.1.6 Distinctions between narrow, general and super-intelligent AI

Weak or narrow AI is AI focused on a specific task (Russell and Norvig, 2016). Popular ML of today, the capability we find on cloud digital services, is narrow AI, or narrow ML, focused on a specific task (e.g. supervised and unsupervised learning). Examples of narrow ML are support vector machines (SVMs), decision trees and K-nearest neighbour (KNN) algorithms. They test a hypothesis based on a specific task. The learning from experience is on a specific, focused task.

Artificial general intelligence (AGI) is a machine capable of learning any intellectual task that a human can. It is hypothetical and not required for AI and ML in general. This hypothetical type of AI is often the subject of speculation and sometimes feared by humans. We are nowhere near AGI, and it could be decades or even centuries before we are close to achieving it.

AGI can be taken one step further. Science fiction is fond of machines that are assumed to have consciousness; strong AI is AGI that is also conscious. Consciousness is complex and difficult. Some consider it to be the hardest problem in AI. This is covered in more detail in Chapter 6, particularly Section 6.4.1.

AGI and strong, or conscious, AGI is not currently feasible and is not realistic in the foreseeable future. It is an active area of research, but not a requirement for AI or ML.

> The origins of consciousness are not known; we can only speculate as to what consciousness is. Professor David Chalmers has posed two questions of consciousness, the easy question and the hard question (Chalmers, 1996):
>
> - The easy question is to explain the ability to discriminate, react to stimuli, integrate information, report mental states – and we can study these with the scientific method.
>
> - The hard question is about how and why physical processes give rise to experience. Why do these processes take place 'in the dark', without any accompanying state of experience?
>
> An essential part of being human is subjective experience. Understanding this conscious experience is not going to be easy. Professor Chalmers and Professor Roger Penrose both think that the answer to the hard question could be as hard as quantum mechanics. Professor Penrose has proposed a mechanism or foundation for consciousness based on quantum mechanics (Penrose et al., 2011). More detail of these concepts can be found in the Further reading list at the end of this book.

There has been considerable debate about consciousness. It is an open question and an academic exercise. One of the core complete rejections of strong AI, at least in digital computers, is presented by Professor John Searle. He argues, using the Chinese Room thought experiment, that strong AI cannot occur, and states (Searle, 2002):

> Computational models of consciousness are not sufficient by themselves for consciousness. The computational model for consciousness stands to consciousness in the same way the computational model of anything stands to the domain being modelled. Nobody supposes that the computational model of rainstorms in London will leave us all wet. But they make the mistake of supposing that the computational model of consciousness is somehow conscious. It is the same mistake in both cases.

Just like Sir James Lighthill's report on AGI that started the AI 'winter' of the 1970s and 1980s (Lighthill, 1973), Professor Searle has given us an argument against the next step for strong AGI as well. In both cases – AGI and strong AI – we should note that they are a long way off and currently narrow, or weak, AI is here and working. Neither AGI nor strong AI are a requirement for AI. We, as humans, set the goals for our AI to achieve. Working within the European Union (EU) ethical guidelines for AI points us in the right direction for our AI to have a human-centric ethical purpose, to be trustworthy and technically robust. This approach echoes the work of Professor Max Tegmark, who emphasises the necessity of goals when making explicit what goals machines will achieve (Tegmark, 2018). What this means is that any explicit AI machine goal should be focused on achieving goals that are aligned with human goals. He shows that we need machines to align with our goals.

1.1.2 Key milestones in AI development

Today we are mainly concerned with the current use and future applications of AI and the benefits we would like to obtain from its use, but we should not ignore where and when AI first appeared, the history of AI and the challenges encountered along the way. To ignore our AI history would be to risk some of those challenges recurring.

Before we look at AI's history, it's worth noting that it is now generally recognised that John McCarthy coined the term 'artificial intelligence' in 1955.

Way back in antiquity there were legends, stories and myths of effigies and artificial mechanical bodies endowed with some form of intelligence, typically the creation of a wise man or master craftsman. We still have references to golems of biblical times, which were magical creatures made of mud or clay and brought to 'life' through some incantation or magic spell. Aristotle (384–322 BCE), the Greek polymath and father of later Western philosophy, was the first to write about objects and logic and laid the foundations of ontology and the scientific method. As a result, today we teach natural science, data science, computer science and social science. Without the required technology, many centuries passed without any progress in the pursuit of AI.

In the 18th century we saw the mathematical development of statistics (Bayes' theorem) and the first computer description and algorithm from Ada Lovelace. During the 19th century, AI entered the world of science fiction literature with Mary Shelley's *Frankenstein* in 1818 and Samuel Butler's novel *Erewhon* in 1872, which drew on an earlier (1863) letter he had written to *The Press* newspaper in New Zealand, 'Darwin among the Machines'. Butler was the first to write about the possibility that machines might develop consciousness by natural selection.

In 1920 Czech-born Karel Čapek introduced the word 'robot' to the world within his stage play, *R.U.R.* (Rossum's Universal Robots). 'Robot' comes from the Slavic language word *robota*, meaning forced labourer.

AI has since become a recurrent theme in science fiction writing and films, whether utopian, emphasising the potential benefits, or dystopian, focusing on negative effects such as replacing humans as the dominant race with self-replicating intelligent machines. To some extent, we can say that what was yesterday's science fiction is quickly becoming today's science fact.

In 1943 Warren McCulloch and Walter Pitts created a computational model for NNs, which opened up the subject. The first was an electronic analogue NN built by Marvin Minksy. This sparked the long-lasting relationship between AI and engineering control theory. Then, in the 1950s, the English mathematician Alan Turing published a paper entitled 'Computing machinery and intelligence' in the journal *Mind* (Turing, 1950). This really opened the door to the field that would be called artificial intelligence. It took a further six years, however, before the scientific community adopted the term 'artificial intelligence'.

John McCarthy organised the first academic conference on the subject of AI at Dartmouth College, New Hampshire, in the summer of 1956. The 'summer school' lasted eight weeks and brainstormed the area of AI. It was originally planned to be attended by

10 people; in fact, people 'dropped in' for various sessions and the final list numbered nearly 50 participants. Russell and Norvig quoted McCarthy's proposal for the summer school (Russell and Norvig, 2016: 17):

> We propose that a 2-month, 10-man study of artificial intelligence be carried out during the summer of 1956 at Dartmouth College in Hanover, New Hampshire. The study is to proceed on the basis of the conjecture that every aspect of learning or any other feature of intelligence can in principle be so precisely described that a machine can be made to simulate it. An attempt will be made to find how to make machines use language, form abstractions and concepts, solve kinds of problems now reserved for humans, and improve themselves. We think that a significant advance can be made in one or more of these problems if a carefully selected group of scientists work on it together for a summer.

The mid-1950s into the early 1960s also saw the start of machines playing draughts, checkers and chess. This was the start of 'Game AI', which is still big business to this day and has developed into a multi-billion-dollar industry. In 1959, Arthur Lee Samuel first used and popularised the term 'machine learning', although today Tom Mitchell's definition is more widely quoted (see Section 1.1.1.4). Samuel's checkers-playing program was among the world's first successful self-learning programs, and was amazing given the limited technology available to him at the time.

The 1960s saw the development of logic-based computing and the development of programming languages such as Prolog. Previously it took an astronomical number of calculations to prove simple theorems using this logic method. It opened up a debate regarding how people and computers think. Hubert Dreyfus, a professor of philosophy at the University of California Berkeley (1968–2017), challenged the field of AI in a book explaining that human beings rarely used logic when they solved problems (Dreyfus, 1972). McCarthy was critical of this argument and the association of what people do as being irrelevant to the field of AI (McCarthy, 2000). He argued that what was really needed were machines that could solve problems – not machines that think as people think.

In 1973 many funding sources for AI research were withdrawn, largely due to Sir James Lighthill's report on the state of AI (Lighthill, 1973), along with increasing pressure from the US Congress and both the US and British governments. Consequently, research funding for AI was significantly reduced, and the difficult years that followed would later be known as an 'AI winter', lasting from 1974 to 1980. Sir James highlighted that other sciences could solve typical AI problems; AI would hit combinatorial explosion limits and practical problems would resign AI to solving only trivial toy problems. In his view, general AI could not be achieved and there was no prospect of a general AI robot ever. During the 1970s, however, a number of breakthroughs were made, most notably in 1974 by Ray Kurzweil's company, Kurzweil Computer Products Inc, developing optical character recognition (OCR) and a text-to-speech synthesiser, thus enabling blind people to have a computer read text to them out loud. It was unveiled to the public in 1976 at the National Federation for the Blind, and became commercially available in 1978. Kurzweil subsequently sold the business to Xerox. It is widely considered to be the first AI product, although today we don't associate OCR with AI or ML because it is now routine and we take it for granted, but it was the precursor to reading handwritten text and the development of NLP.

Following on from the 'AI winter', the 1980s saw a boom in AI activity. In 1986 David Rumelhart and James McClelland developed ideas around parallel distributed processing and NN models. Their book, *Parallel Distributed Processing: Explorations in the Microstructure of Cognition* (1986), described their creation of computer simulations of perception, giving computer scientists their first testable models of neural processing. The 1980s also saw the rise of the robots, with many researchers suggesting that AI must have a body if it is to be of use; it needs to perceive, move, survive and deal with the world. This led to developments in sensor-motor skills development. AI also began to be used for logistics, data mining, medical diagnosis and in other areas.

During the 1990s a new paradigm called 'intelligent agents' became widely accepted in the AI community. An intelligent agent is a system that perceives its environment and takes actions that maximise its chances of success. In 1997, IBM's Deep Blue became the first computer chess-playing system to beat a reigning world chess champion, Garry Kasparov. It won by searching 200,000,000 moves per second. By comparison, 20 years on, Apple's iPhone7 was 40 times faster than Deep Blue had been in 1997. We have supercomputers in our pockets.

1.1.2.1 Bringing us up to date, the last 20 years...
Over the last 20 years, developments in technology with cheaper and faster computers have finally caught up with our AI aspirations. We have started to gain the computing power to really put AI to work.

In 2002 iRobot released Roomba, which autonomously vacuums a floor while navigating and avoiding obstacles. It sold a million units by 2004, and over 8 million units by 2020. iRobot then went on to create a range of other commercial, environmental, military and medical robots. Today, you'll see garden lawns maintained by AI lawn mowers, feeders and sprinklers.

In 2004 the Defense Advanced Research Projects Agency (DARPA), a prominent research organisation of the United States Department of Defense, introduced the DARPA Grand Challenge, offering prize money for competitors to produce vehicles capable of travelling autonomously over 150 miles. Then, in 2007, DARPA launched the Urban Challenge for autonomous cars to obey traffic rules and operate in an urban environment, covering 60 miles within six hours.

Google entered the self-driving autonomous market in 2009 and built its first autonomous car, which sparked a commercial battle between Tesla, General Motors, Volkswagen and Ford, to name a few entrants into the same market.

From 2011 to 2014 a series of smartphone apps were released that use natural language to answer questions, make recommendations and perform actions: Apple's Siri (2011), Google's Google Now (2012) and Microsoft's Cortana (2014).

SCHAFT Inc of Japan, a subsidiary of Google, built robot HRP-2, which defeated 15 teams to win DARPA's Robotics Challenge Trial in 2013. The HRP-2 robot scored 27 out of 32 points over eight tasks needed in disaster response. The tasks were to drive a vehicle, walk over debris, climb a ladder, remove debris, walk through doors, cut through a wall, close valves and connect a hose.

In 2015 an open letter petitioning for the ban in development and use of autonomous weapons was signed by leading figures such as Stephen Hawking, Elon Musk, Steve Wozniak and over 3,000 researchers in AI and robotics.

In 2016 Google's DeepMind AlphaGo supercomputer beat Lee Se-dol, a world Go champion, at a five-game match of Go (a strategic board game that originated in China more than 2,500 years ago and considered one of the most complex strategy games in the world). It took just 30 hours of unsupervised learning for the supercomputer to teach itself to play Go. Lee Se-dol was a ninth dan professional Korean Go champion who won 27 major tournaments from 2002 to 2016. He announced his retirement from the game in 2019, declaring that AI has created an opponent that 'cannot be defeated' (Vincent, 2019). In 2017 AlphaGo Zero, an improved version of AlphaGo, beat the world's best chess-playing computer program, StockFish 8, winning 28 of the 100 games and drawing 72 of them. What is astonishing is that AlphaGo Zero taught itself how to play chess in under four hours.

Also in 2017, the Asilomar Conference on Beneficial AI was held near Monterey, California (Wikipedia Contributors, 2024j). Thought leaders in economics, law, ethics and philosophy spent five days in discussions dedicated to beneficial AI. It discussed AI ethics and how to bring about beneficial AI while at the same time avoiding the existential risk from AGI.

Figure 1.1 shows a simple schematic of how AI and ML have evolved. Here, in 2024, we see the emergence of foundation models giving AI its toolkit to build applications and products that harness models of vast amounts of data. All the major tech companies are producing foundation models for professionals to harness. AI users now have generative AI at their fingertips. Foundation models are a general description of multipurpose transformer models. Rothman uses the foundation models term to emphasise the vast array available to us and describes them as a paradigm shift (Rothman, 2024). To put this into real terms, the obvious example is OpenAI's ChatGPT.

Figure 1.1 Where AI sits

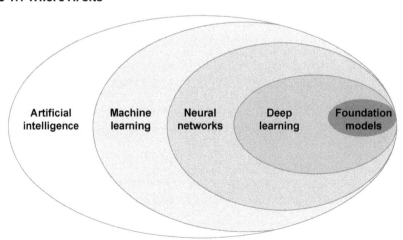

1.1.3 Types of AI

In this section we look at the broader descriptions of this universal subject AI. We may come across several types based on functionality, capability and application. We can think of AI types as versions on a theme or a means to end.

- **Functionality types** include reactive machines, limited memory, theory of mind and self-aware AI.
- **Capability types** include narrow AI (or weak AI), general AI (or strong AI) and super-intelligent AI.
- **Application types** include expert systems, NLP, computer vision, robotics and recommendation systems.

This list is not exhaustive because the pace of development is opening up new creative and exciting directions at a highly motivating and ever increasing speed. Today, we have access to narrow AI that is focused on a well-defined or engineered specific task.

1.1.4 The industrial revolutions

We are currently in the middle of the fourth industrial revolution, and some would argue that we have already reached the fifth, although that has yet to be defined. In each industrial revolution, mankind has designed and developed technologies that have made a paradigm shift in human capabilities and exploited those technologies to drive through progress, although some would argue that not all of mankind benefited as a result of these advancements.

The first industrial revolution occurred during the 18th and 19th centuries, primarily in Europe and the United States. It mainly grew between 1760 and 1840 (although the exact dates are open to debate) and resulted in the transition from hand production methods to machine production – initially led by the development of the steam engine powering large textile factories. It was a major turning point in history, led to worldwide trading and rapid population growth and resulted in large swaths of rural societies becoming urbanised and industrial.

The second industrial revolution occurred between 1870 and 1914, primarily across Europe, the United States and Japan. Sometimes known as the Technological Revolution, again it was a paradigm shift for mankind with the introduction of mass production and assembly lines. Increased use of electricity allowed for advancements in manufacturing and production, and resulted in technological advances such as the internal combustion engine, the telephone and the light bulb.

The third industrial revolution started in the 1950s and is often known as the Digital Revolution. It brought us space exploration, biotechnology, semiconductors, mainframe computing and information and communications technology (ICT), and embedded technology into society with personal computers, the internet and automated production of goods.

There is a bit of a blur between the third and fourth industrial revolutions. We believe that we are in the fourth today. The fourth exploits the gains made in the digital revolution and is disruptive, driven by AI, robotics, Internet of Things (IoT), three-dimensional (3D) plastic printing, nanotechnology, bioengineering and so on.

I know what you're thinking: 'What happened between the industrial revolutions?' We didn't stop inventing or improving between the revolutions, and different parts of the globe experienced them at different times at different speeds, but these particular periods were paradigm shifts in thinking and invention.

The Diffusion of Innovation theory, developed by E.M. Rogers in 1962, explains how, over time, an idea or product gains momentum and spreads (diffuses) through a specific population, and he classifies adopters of innovations into five adopter categories: innovators, early adopters, early majority, late majority and laggards. This is based on the idea that certain individuals are inevitably more open to adoption and adaption than others. The adoption of AI technologies globally will be faster than the adoption of other technologies because we have seen a reduction in the time an industrial revolution lasts, from centuries to decades and now fractions of a decade, but we will still be led by innovators and there will still be laggards and even Luddites.

Human intelligence has led us through the various industrial revolutions. We are in the middle of the fourth, but what does this mean if we think about AI? It means increasingly that we will have more robots doing routine, monotonous, laborious and dangerous tasks – doing the 'heavy lifting'. This introduces the idea of humans and machines working together at what they are each good at.

1.2 WHAT ARE ETHICS AND TRUSTWORTHY AI?

AI is part of the ethics of how we operate as individuals, as organisations and in the wider context of the universe. At its core are our ethical principles, rights and values. Ethics brings unique challenges for philosophers and academics, in particular the ethical implications of AI challenging our freewill or agency, perhaps even having access to conscious capabilities. We'll explore what this means in more detail in Chapter 2. The past decade has seen a movement towards trustworthy AI, capturing the spirit of Ada Lovelace (Wikipedia Contributors, 2024a), who questioned the implications of technology on society, and more recently academic works from Stuart Russell and AI's human compatibility (Russell, 2019) and Max Tegmark's idea in *Life 3.0* (2018) to let humans set the goals as well as designing the AI fire extinguisher before we start the AI fire. The culmination of this pioneering work has seen progress towards the EU's law, The Artificial Intelligence Act, and AI having a human-centric ethical purpose that is technically robust and trustworthy (European Parliament, 2023). International standards give us a broad range of literature to draw on, and we'll add more detail on this in Chapter 2.

The world of AI is constantly changing, and the rate of change is not linear; you may not have noticed, but it is actually growing exponentially, which basically means it is accelerating faster and faster on an annual basis. Some would argue that its growth is potentially out of control and we need to put the brakes on (Future of Life, 2023); others suggest that we just let it run and see where it goes.

Many people have fears that not all AI improvements are necessarily beneficial to humankind or in society's best interests. For example, there is the potential to weaponise AI, for it to intrude into our lives and privacy by listening and watching everything, and for individuals' identification without consent with covert AI systems. There is also the potential to have control of our personal data taken away from us. So, how should AI be managed and controlled? Or, what happens if we try to do the right thing – whatever that means – and we get it wrong?

On a day-to-day practical level, we are using narrow AI that works on a well-engineered or well-defined task.

1.2.1 AI as part of universal design

The concept of universal design was coined by the architect Ronald Mace, and is the design and composition of an environment so that it can be accessed, understood and used to the greatest extent possible by all people. For example, door handles, elevator controls and light switches should be designed for use by all people regardless of their age, size, ability or disability. A consideration, therefore, for any new AI service, system or product is that it is designed for all.

Incorporating the potential of AI in universal design can allow someone who is visually impaired to ask a 'home assistant' what the weather is like, or for someone who is physically incapacitated to turn on the heating, or someone who is travelling home to turn on the heating while travelling. This is an exciting area where AI can bring superhuman capabilities to everyone.

A human working alongside an intelligent AI-enabled machine has the capability to do a lot more, whether that is in a work environment reducing physical risk or exertion, or on a personal basis educating us or translating our conversations. AI systems have the potential to make us more human.

Our efforts and endeavours developing AI-enabled products, systems and services should be focused on allowing us to be more human, improving us as humans (improving our physical and mental performance or by making us more active) or improving our ability to communicate or socialise.

The continual emergence of AI systems and products means that we as individuals and as part of wider society are going to have to reimagine every area of our lives to use AI in a positive way for all.

1.2.2 The impact of AI on society

How will AI change our society? We need to consider what implications AI will have on wider society now and in the long term, and what freedoms and human values we are prepared to give up in return for the benefits we will enjoy. Increased use of AI and wider adoption will come with some risks and challenges, such as the potential of its weaponisation, but it may also solve climate change and world poverty. How can we balance the risks versus the benefits?

As a society, we need to start preparing now for how we manage AI today and in the future. For example, would you be prepared to give up your job to an AI entity but in return receive a basic universal income? Maybe, but would you be prepared to give up your child's future career dreams and aspirations for short-term gains here and now? What happens if we charge off with the aim of utopia and get it wrong?

We have previously talked about how AI will affect us as individuals in our workplace by improving efficiencies and augmenting what we do and are capable of. Of course, our human-centric ethical purpose can be used to mitigate disadvantages that affect us in our everyday lives, removing inequalities and allowing us to become more human, undertaking tasks that involve creativity and empathy, among others, in particular, helping to remove bias and improve sustainability and equality of outcome.

1.2.3 Sustainability and environmental impact

Intergenerational equity is the concept or idea of fairness or justice in relationships between generations. It is often considered as a value concept that focuses on the rights of future generations. It describes how each future generation has the right to inherit the same level of diversity in natural and cultural resources enjoyed by previous generations, and equitable access to the use and benefits of these resources. A common example often quoted is current generations running up debt that will have to be paid off by future generations to come.

The study of intergenerational equity and view of sustainability is based on three pillars: economic, social and environmental (see Figure 1.2).

- Economic – many economists have tried to predict the effect that AI will have on the economy. PwC forecast in 2017 that AI could contribute up to US$15.7 trillion to the global economy in 2030 (Rao and Verweij, 2017), more than the current output of China and India combined. Of this, US$6.6 trillion is likely to come from increased productivity and US$9.1 trillion is likely to come from consumption-side effects.

- Social – during the 1970s in the UK, whole communities felt the impact of large organisations closing down operations, including airlines, dockyards and mines. AI brings with it big headlines of major disruption and potential job losses. In an era of humans and machines, the possibility exists for humans to move onto higher value work or to enrich and develop their talents while machines take away the heavy lifting, reducing the burden on human effort. Sustainability is part of the EU's AI guidelines, in particular the human-centric ethical purpose. It is essential, therefore, that assessing the impact of AI on society is a part of that.

- Environmental – the AI carbon footprint most certainly has an environmental impact, which needs to be factored into its business case. Donna Lu suggested in a 2019 article in *New Scientist* that 'Creating an AI can be five times worse for the planet than a car' (Lu, 2019).

Figure 1.2 The three pillars of sustainability

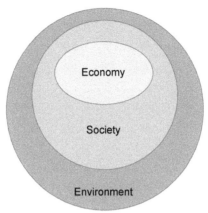

1.3 SUMMARY

This chapter has introduced the basic concepts that underlie AI and its relationship to human intelligence. The chapter sets the stage for understanding AI by framing it as a learning agent, similar to human intelligence, that interacts with its environment to achieve goals. AI encompasses creating intelligent entities, while ML involves the ability of systems to learn from data.

The chapter has laid the groundwork for a deeper exploration of AI, clarifying its goals and limitations. By understanding the key principles of intelligence, learning and the scientific method, you are prepared for a more detailed analysis of AI's applications, ethical implications and future developments in subsequent chapters.

2 ETHICS AND GOVERNANCE

In this chapter we will make sense of how ethics, and how AI ethics, makes us think more carefully about the benefits of if and how AI can improve things – a human-centric ethical purpose. This is essentially what humans value, which is different for each person based on the context, and changes with time. We may also think of this as our principles that lead to the values and rights we expect.

AI and ML are by no means a panacea – a silver bullet. Fred Brooks helped us understand complexity by distinguishing between two types: accidental complexity, which can be fixed, and essential complexity, which must be acknowledged and addressed by the system itself (Brooks, 1986). Brook's careful thought gives us a warning that when we engineer accidental complexity there is a fair chance that AI/ML can help us, but essential complexity is something that AI/ML may not help us with. Making sense of a problem we face can be further enhanced by frameworks such as Cynefin (Cynefin, 2023). Our team, including the subject matter experts, must work with stakeholders to ensure ethics and governance is built in from the start and for the full life cycle of the project. This means our AI team integrates a human-centric ethical purpose into the design of AI systems, guided by our ethical principles (Floridi and Cowls, 2019), which prioritise protecting human rights and values.

2.1 ETHICAL AI

As part of the EU's guidelines on ethical AI, we must understand the ethical challenges AI raises (European Commission, 2024). This is by no means a simple subject; we must consider, for example:

- How do we deal with the ethics?
- Who is responsible?
- What does the law say?
- What about human rights?
- What about machine rights?

With AGI and strong AI, robotics becomes another ethical question. General approaches to robotic AI have been published by the Engineering and Physical Sciences Research Council (Wikipedia Contributors, 2024k). Other organisations are preparing AI principles also; these include the International Organization for Standardization (ISO). The output from organisations like the ISO include standards on definitions (ISO/IEC, 2022a), risks (ISO/IEC, 2023b), management (ISO/IEC, 2023a) and guidance (ISO/IEC, 2022b).

Key themes coming out from various organisations include transparency, accountability and care taken when considering bias, diversity, inequalities and unemployment. The themes link back to good ethics. The EU has built the guidelines for AI around an ethical purpose that we can measure (European Commission, 2024) using, for example, the United Nations sustainability goals (United Nations, 2015). These guidelines have now become AI law in the EU (European Parliament, 2023).

2.1.1 AI governance

AI governance refers to the framework of policies, processes and controls designed to ensure the ethical, transparent and accountable development and deployment of AI systems. It involves setting standards for data usage, algorithmic fairness and decision-making processes to mitigate risks such as bias, privacy violations and lack of explainability. Effective AI governance aims to balance innovation with the protection of individual rights and societal values, fostering trust and compliance with legal and ethical standards.

2.1.2 Human-centric ethical purpose – fundamental rights, principles and values

A human-centric ethical purpose aims to enhance human capabilities and societal wellbeing. It builds on the Universal Declaration of Human Rights and the Charter of Fundamental Rights of the European Union. Typical measures of how this might be demonstrated are the United Nations sustainability goals. There are several bodies and many organisations worldwide currently working on codes of ethics around AI. We should also note in passing that the Center for Human-Compatible Artificial Intelligence (https://humancompatible.ai) is also producing papers on AI and its human-centric use. This centre was founded by the author of *Human Compatible: Artificial Intelligence and the Problem of Control*, Stuart Russell (2019) – the book that addresses the control problem of AI head on.

2.1.2.1 Future of Life Institute (US founded)
The Future of Life Institute has developed the Asilomar Principles for AI (https://futureoflife.org). There are 23 principles relating to:

- Research – goals, funding, policy, cultures, race avoidance (speed of progress).
- Ethics and values – safety, failure transparency, judicial transparency, responsibility, value alignment, human values, personal privacy, liberty and privacy, shared benefit, shared prosperity, human control, non-subversion, AI arms race.
- Longer term issues – capability caution, importance, risks, recursive self-improvement, common good.

The name Asilomar was inspired by the highly influential Asilomar Conference on Recombinant DNA in 1975. The biotechnology community are still influenced by the voluntary guidelines developed at this conference.

2.1.2.2 The European Union – AI guidelines and AI Act

The EU is one of the trailblazers tackling ethical use of AI from the front. It has produced AI guidelines that have been made fundamental and explicit in its AI Act.

AI ethics guidelines are produced by the European Commission's High-Level Expert Group on Artificial Intelligence (European Commission, 2024). The guidelines are for trustworthy AI.

Trustworthy AI basically has two components:

1. It should respect fundamental rights, applicable regulation and core principles and values, ensuring an 'ethical purpose'.
2. It should be technically robust and reliable, since, even with good intentions, a lack of technological mastery can cause unintentional harm.

It also considers:

- Rights – a collection of entitlements that a person may have and that are protected by government and the courts.
- Values – ethical ideals or beliefs for which a person has enduring preference, and which determine our state of mind and act as a motivator.
- Principles – a fundamental well-settled rule of law or standard for good behaviour, or collectively our moral or ethical standards.

The last revision of the EU's 'ethical guidance on AI' was produced in April 2019, and this has now evolved into the European Union Artificial Intelligence Act 2024 (European Parliament, 2023).

The EU AI Act, proposed by the European Commission, represents a pioneering regulatory framework aimed at governing AI technologies within the EU. It seeks to ensure that AI systems are developed and utilised in a manner that is ethical, transparent and aligned with fundamental rights. The Act categorises AI applications into risk-based tiers, ranging from minimal to high risk, with stringent requirements imposed on the latter. These include rigorous conformity assessments, data governance standards and continuous monitoring to mitigate potential harms. By setting these comprehensive standards, the AI Act aspires to foster innovation while safeguarding public trust and safety, positioning the EU as a global leader in the responsible deployment of AI.

2.1.2.3 The ISO/IEC

ISO/IEC stands for the International Organization for Standardization (ISO) and the International Electrotechnical Commission (IEC). These two organisations collaborate to develop and publish international standards in various fields, including information technology and telecommunications. ISO focuses on a broad range of industries and services, while IEC specialises in electrical, electronic and related technologies. Together, they produce joint standards to ensure quality, safety, efficiency and interoperability in technological and industrial sectors globally.

The ISO/IEC standards are an international effort to make our AI human-centric ethical purpose more explicit. These are explained in more detail Section 2.2.3.

2.1.2.4 Japan, USA, Canadian and UK AI bodies

Internationally, Japan has also adopted the human-centric approach to AI. It published the 'Social principles of human-centric AI' (Cabinet Office, Government of Japan, 2022). The National Institute of Standards (NIST) are active producing US guidelines and standards (NIST, 2023). In addition, The White House has published the 'Blueprint for an AI bill of rights' (The White House, 2022). Their neighbours, the Canadians, are at the heart of the ISO/IEC's standard 42001 (ISO/IEC, 2023a) and also published on the responsible use of AI in government (Government of Canada, 2024). The UK's Alan Turing Institute has published an AI standards hub, a sensible starting point for AI publications, standards and ethics (The Alan Turing Institute, 2024).

2.2 UNDERSTANDING THE RISKS AI BRINGS TO PROJECTS

The challenges to the development and implementation of AI solutions can manifest themselves in several ways, including:

- self-interest;
- self-review;
- conflict of interest;
- intimidation;
- advocacy.

These challenges are not trivial and will require managing. Of particular importance is the false advocacy of AI's capabilities; some might say this led to the AI winters. Ethical challenges can be compromised by intimidation (think of adversarial debate between politicians). Conflicts of interest or self-interest must be managed at all times, but the use of AI introduces the need for transparency, and reviews and audits of the AI's output are necessary to ensure it is fit for purpose and fair.

2.2.1 Ethical concerns in AI

We all want beneficial AI systems that we can trust, so the achievement of trustworthy AI draws heavily on the field of ethics. AI ethics could be considered a subfield of applied ethics and technology, and focuses on the ethical issues raised by the design, development, implementation and use of AI technologies.

The goal of AI ethics is therefore to identify how AI can advance or raise concerns about the 'good' life of individuals, whether this be in terms of quality of life, mental autonomy or freedom to live in a democratic society. It concerns itself with issues of diversity and inclusion (with regard to training data and the ends to which AI serves) as well as issues of distributive justice (who will benefit from AI and who will not).

What comes out of consideration of AI ethics are some general themes that most agree on:

- **Transparency** – we might think of this as how we understand what went wrong when AI gets it wrong. Can we be explicit as to why an AI technology failed? For instance, a NN is generated by an algorithm that we can understand; however, for it to be transparent we would also need to understand explicitly why the NN came to a particular outcome. It gives us a basis on which to learn how the AI system performed and why.

- **Accountability** – we need to be careful or we could get inequality, bias, unemployment and so on.

- Weaponisation – creating lethal autonomous weapons is a red line that we must not cross; however, national security is a special subject that is beyond the general use of AI in society.

- **Harm** – the AI must do no harm.

- **Human dignity** – protecting an individual's right to be valued and respected for their own sake, treated them ethically.

- **Safety and security** – almost goes without saying, but this is an important ethical concern.

- **Privacy** – this has been a source of much political debate and is now a fundamental requirement, for example with the implementation of the General Data Protection Regulation (GDPR) in the EU.

- **Sustainability** – a set of goals that align with our stakeholders and the wider community, enshrining intergenerational equity.

- **Fairness** – along with sustainability, bias raises more ethical challenges we must embrace and manage well.

- **Social benefit** – how does everyone benefit from the opportunities of AI?

2.2.2 Guiding principles for ethical AI

Ethics is a philosophical subject that manifests itself in individuals as what we term our moral principles; in fact, many people use the words interchangeably. Basically, both ethics and morals relate to 'right' and 'wrong' conduct.

It is worth noting that ethics, the subject that academics and students study, is treated as singular, for example 'my elected subject this term is ethics'. Our moral principles or ethical principles that motivate and guide us are treated as plural. This way we can differentiate between the subject, which is singular, and a person's ethical principles, which are plural.

Further expansion on ethics can be found in *An Intelligent Person's Guide to Ethics* by Lady Warnock (2006).

Morals tend to refer to an individual's own personal principles regarding what is right and wrong. Ethics generally refer to rules provided by an external source or community, for example codes of conduct in the workplace or an agreed code of ethics for a profession such as medicine or law.

Ethics as a field of study is centuries old and centres on questions such as:

- What is a 'good' action?
- What is 'right'?
- What is a 'good' life?

2.2.3 Addressing ethical challenges

Addressing ethical challenges not only gives us an abundance of knowledge to work with, but it also addresses the rogue actors who do not provide AI services and products ethically. It builds on the Human Rights Act and data protection legislation. Specific legal frameworks give us the means to enforce ethical AI, such as the EU's AI Act. So where do we start?

Figure 2.1 shows how we can gather guidance from various sources to help us develop a human-centric AI that is both trustworthy and robust, with a strong ethical foundation. We can start with the Alan Turing Institute's AI Standards Hub (2024) and the ISO/IEC's 42001 standard (ISO/IEC, 2023a). From here we might adopt the EU's AI guidelines to produce a set of requirements. This is shown schematically in Figure 2.2. The schematic identifies that the 17 United Nations' sustainability goals can be interwoven into our ethical purpose and also that the AI project is iterative where we are always learning from experience.

___

Figure 2.1 Practical AI ethics

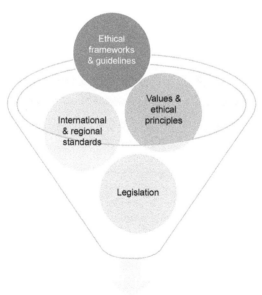

Practical AI ethics

ISO/IEC 42001 is considered a good AI governance model as it provides comprehensive guidelines for establishing, implementing and maintaining AI management systems.

It emphasises risk management, ethical considerations, transparency and continuous improvement. By covering all phases of the AI lifecycle and extending controls to third-party suppliers, it ensures responsible AI development and deployment. Adopting ISO/IEC 42001 helps organisations to build trustworthy AI systems, align with regulatory requirements and promote ethical AI practices.

ISO/IEC 42001 is an artificial intelligence management system (AIMS) that can be used within organisations. This standard aims to ensure responsible development, deployment and use of AI technologies by addressing key areas such as risk management, AI impact assessments, system lifecycle management and supplier management. By adhering to ISO/IEC 42001, organisations can build trustworthy AI systems, enhance governance and promote ethical practices, thereby fostering trust and compliance with upcoming regulations such as the EU AI Act.

Figure 2.2 Technical and non-technical methods to achieve trustworthy AI (EU Creative Commons Attribution 4.0 International (CC BY 4.0) licence)

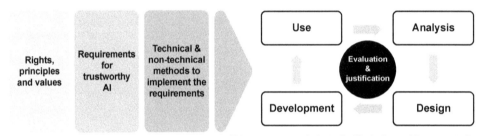

"AI systems can help to facilitate the achievement of the UN's Sustainable Development Goals, such as promoting gender balance and tackling climate change, rationalising our use of natural resources, enhancing our health, mobility and production processes, and supporting how we monitor progress against sustainability and social cohesion indicators."

2.3 REGULATION AND RISK MANAGEMENT

Up to June 2024, there hasn't been significant regulation of AI apart from the EU AI Act. Regulating AI is crucial to ensure safety, fairness and accountability. It helps to mitigate risks such as bias, discrimination and privacy violations, while promoting ethical standards and transparency. Regulations can also prevent harmful applications, such as autonomous weapons, and ensure that AI development aligns with societal values and human rights. By setting clear guidelines and oversight, regulations foster public trust in AI technologies and facilitate their responsible and beneficial use.

The Web Content Accessibility Guidelines (WCAG) regulate the use of AI by providing standards to ensure web content, including AI-driven interfaces, is accessible to people with disabilities (www.w3.org/TR/WCAG21/).

WCAG focuses on making content perceivable, operable, understandable and robust for all users. This includes guidelines for AI systems to offer alternative text for images, ensure navigability with assistive technologies, provide clear and understandable language and maintain compatibility with various devices and user agents.

These standards help to ensure AI technologies are inclusive and accessible to everyone.

The regulation and risk management should form part of a unified strategy based on ISO 31000 (ISO/IEC, 2018) and be supplemented by ISO/IEC 23894 (ISO/IEC, 2023b), or, perhaps, NIST's risk management framework (NIST, 2024) supplemented by the AI risk management framework (NIST, 2023). Both NIST and ISO approaches make risk assessment, governance, transparency, bias mitigation, security and privacy, human oversight and continuous monitoring fundamental to the management of risk with AI.

To assess the risk, the EU's AI Act (European Parliament, 2023) provides a useful way to assess the risks of an AI project, product or service. This is shown in Figure 2.3.

We can see some important categories: unacceptable risk, high risk, transparency risk and minimal risk. The EU AI Act is a landmark legislative proposal aimed at regulating the development, deployment and use of AI within the EU. It is the world's first comprehensive AI legislation. It seeks to ensure that AI technologies are safe, transparent and aligned with fundamental rights and values.

2.3.1 Role of regulation in AI

Here are the key objectives of the EU AI Act:

- Ensure the safety and fundamental rights of people and businesses while fostering AI innovation.
- Strengthen trust in AI by providing clear rules and guidelines.
- Promote investment and innovation in AI across the EU.

Unacceptable risk AI systems are systems considered a threat to people and will be banned. They include:

- The cognitive behavioural manipulation of people or specific vulnerable groups, for example voice-activated toys that encourage dangerous behaviour in children.
- Social scoring, which is classifying people based on behaviour, socio-economic status or personal characteristics.
- The biometric identification and categorisation of people.
- The real-time and remote biometric identification systems, such as facial recognition.

Figure 2.3 The EU AI Act risk pyramid

In April 2021, the European Commission proposed the first EU regulatory framework for AI. It says that AI systems that can be used in different applications are analysed and classified according to the risk they pose to users. The different risk levels will mean more or less regulation.

The new rules establish obligations for providers and users depending on the level of risk from artificial intelligence. While many AI systems pose minimal risk, they need to be assessed.

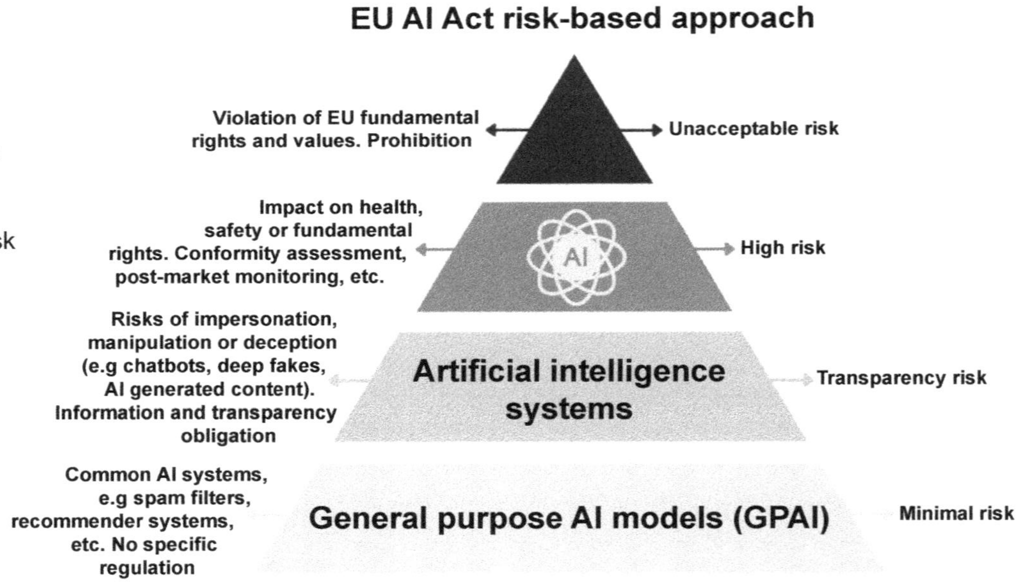

EU AI Act risk-based approach

Unacceptable risk

Violation of EU fundamental rights and values. Prohibition

High risk

Impact on health, safety or fundamental rights. Conformity assessment, post-market monitoring, etc.

Artificial intelligence systems

Transparency risk

Risks of impersonation, manipulation or deception (e.g chatbots, deep fakes, AI generated content). Information and transparency obligation

General purpose AI models (GPAI)

Minimal risk

Common AI systems, e.g spam filters, recommender systems, etc. No specific regulation

AI systems that negatively affect safety or fundamental rights will be considered high risk and will be divided into two categories:

1. AI systems that are used in products falling under the EU's product safety legislation. This includes toys, aviation, cars, medical devices and lifts.

2. AI systems falling into specific areas that will have to be registered in an EU database:

   - Management and operation of critical infrastructure.

   - Education and vocational training.

   - Employment, worker management and access to self-employment.

   - Access to and enjoyment of essential private services and public services and benefits.

   - Law enforcement.

   - Migration, asylum and border control management.

   - Assistance in legal interpretation and application of the law.

As part of the Act, there are transparency risk requirements: generative AI, such as ChatGPT, will not be classified as high risk, but will have to comply with transparency requirements and EU copyright law:

- disclosing that the content was generated by AI;

- designing the model to prevent it from generating illegal content; and

- publishing summaries of copyrighted data used for training.

We must abide by the law. Regulation is the control of AI by law enforced by the courts. This is different from ethics, in that we cannot be punished by breaking an ethical principle but there are consequences for breaking a law. The first and specific AI legal framework is the EU's AI Act (European Parliament, 2023). This EU regulation applies to anyone who makes, uses, imports or distributes AI systems in the EU, regardless of where they are based. It also applies to AI systems used in the EU, even if they are made outside the EU. Saying that, the legal landscape is changing as quickly as the field of AI is developing, and we can expect more. We must also abide by existing laws such as the EU GDPR (European Commission, 2018) and the UK's Data Protection Act (UK Government, 2018). Abiding by these regulations is a fundamental part of an ML or AI project.

2.3.2 Risk management in AI

Risk is defined as the effect of uncertainty on objectives, which can be positive or negative. It encompasses potential events and their consequences, as well as the likelihood of occurrence (ISO/IEC, 2018).

Risk management is defined as the process of identifying, assessing and controlling threats to an organisation's capital and earnings. These threats, or risks, could stem from various sources including financial uncertainty, legal liabilities, strategic management errors, accidents and natural disasters (ISO/IEC, 2018). Risk analysis is used to identify AI risks and involves systematically assessing the potential threats and vulnerabilities associated with the development, deployment and operation of AI systems. The key steps are:

- Define scope and objectives:
  - Identify assets and processes.
  - Identify threats and vulnerabilities.
  - Assess impact and likelihood.
  - Prioritise risks.
  - Mitigate strategies.
  - Monitor and review.

The risks and challenges associated with AI adoption include:

- bias and discrimination;
- privacy concerns;
- job displacement;
- accountability and transparency;
- regulatory compliance;
- security risks;
- technical limitations;
- ethical dilemmas.

Addressing these challenges requires comprehensive strategies involving ethical guidelines, robust security measures, continuous monitoring and adherence to regulatory frameworks. ISO 31000 provides a framework for risk management that helps organisations to identify, assess and prioritise risks, and develop strategies to manage them effectively.

The standard emphasises:

- **Principles:** establishing a risk management culture that is integral to the organisation.
- **Framework:** building a supportive environment through leadership, commitment and integration into organisational processes.
- **Process:** implementing a systematic approach to identifying, analysing, evaluating, treating, monitoring and reviewing risks.

This approach ensures that risk management is structured, comprehensive and tailored to the organisation's context and objectives.

2.3.2.1 SWOT, PESTLE and Cynefin

In this section we give a brief overview of some techniques to make sense of risk so that we can manage it. The first is SWOT analysis, which looks at the opportunities and threats both internally and externally to your organisation. The second is PESTLE, a process to understand stakeholders. Finally, there's Cynefin, the way to make sense of your AI project and manage complexity and chaos.

By using a SWOT analysis to manage AI risks, organisations can gain insights into their internal capabilities and external environment, enabling them to make informed decisions and take proactive measures to mitigate risks and maximise the benefits of AI adoption. This is shown schematically in Figure 2.4.

Figure 2.4 The SWOT analysis

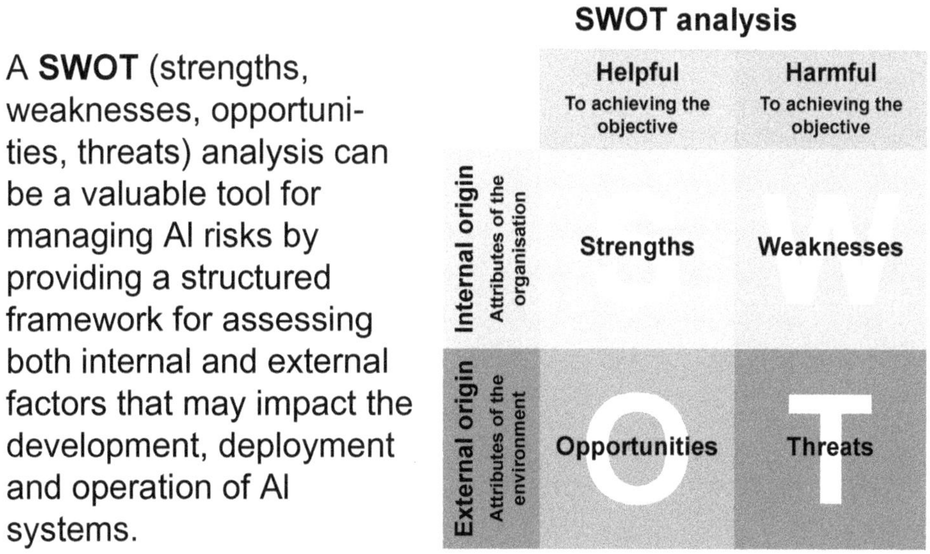

A **SWOT** (strengths, weaknesses, opportunities, threats) analysis can be a valuable tool for managing AI risks by providing a structured framework for assessing both internal and external factors that may impact the development, deployment and operation of AI systems.

SWOT analysis

	Helpful To achieving the objective	Harmful To achieving the objective
Internal origin Attributes of the organisation	Strengths	Weaknesses
External origin Attributes of the environment	Opportunities	Threats

The output from the SWOT analysis, an action plan, can be used to take advantage of opportunities and mitigate risks. The action plan can be used to learn from experience as the project progresses.

With PESTLE analysis, organisations can gain a comprehensive understanding of the external factors influencing AI risks and opportunities. This analysis enables organisations to develop strategic plans and risk management strategies that account for the broader context in which AI operates, ensuring alignment with regulatory requirements, societal expectations and environmental considerations while maximising the benefits of AI adoption. This is shown schematically in Figure 2.5.

Cynefin, shown schematically in Figure 2.6, provides a structured framework for understanding and managing AI risks by acknowledging the complexity and uncertainty inherent in AI systems.

29

Figure 2.5 The PESTLE analysis

A **PESTLE** analysis is a strategic tool used to identify and analyse the external factors that may impact an organisation's operations and decision-making.

PESTLE stands for political, economic, sociological, technological, legal and environmental factors.

When applied to managing AI risks, a PESTLE analysis helps organisations to understand the broader context in which AI operates and the potential risks and opportunities associated with these external factors.

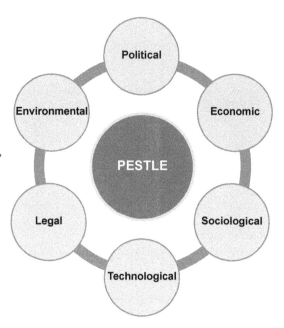

Figure 2.6 Cynefin

Cynefin is a sense-making framework that helps organisations to understand the nature of the challenges they face and make appropriate decisions based on the context.

It categorises situations into five domains: clear, complicated, complex, chaotic and disorder.

Each domain requires a different approach to decision-making and problem-solving.

When it comes to managing AI risks, Cynefin can be a valuable tool for understanding and addressing the complexities involved.

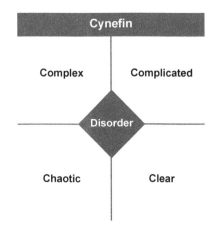

The Cynefin framework was developed by David J. Snowden in 1999. It aims to help leaders understand that every situation is different and requires a unique approach to decision-making. The framework outlines five situational domains that are defined by cause-and-effect relationships. By categorising risks based on their nature and context, organisations can tailor their risk management strategies and responses accordingly, leading to more effective risk mitigation and decision-making in the realm of AI. In addition, Cynefin helps the domain expert and AI experts to assess if narrow AI can help. Narrow AI focused on a well-defined task can be assessed more readily in the clear domain. The complicated domain requires more careful consideration of how narrow AI can help; the domain expert will play a pivotal role in this domain. In the complex domain, the management approach must embrace experimentation, adaptive approaches and scenario planning to navigate uncertainties. It will use techniques such as simulation, modelling and scenario analysis to understand potential risks and develop agile responses as situations evolve. This could involve more sophisticated AI based on domain expert simulations, or, as Brooks described, it will more than likely include essential complexity.

2.3.2.2 Human-centred AI

Human-centred AI (HCAI) is an approach to designing and deploying AI systems with a primary focus on enhancing human capabilities, ensuring meaningful user control and promoting human values. HCAI aims to create AI technologies that are not only technically effective but also socially and ethically responsible.

The key principles of HCAI are:

- **User empowerment:** AI systems should empower users by augmenting their abilities and supporting their goals rather than replacing human roles. The aim is to create tools that enhance human decision-making, creativity and problem-solving.

- **Transparency and explainability:** AI systems should be transparent about how they operate and make decisions. Explainable AI (XAI) techniques are often employed to ensure that users can understand, trust and appropriately manage the AI's outputs.

- **Ethical considerations:** HCAI emphasises the ethical implications of AI, including fairness, accountability and mitigation of biases. It seeks to design AI systems that respect user privacy, ensure data security and promote equity.

- **Usability and user experience (UX):** the design of AI systems should prioritise usability and UX, ensuring that they are intuitive, accessible and responsive to user needs. This includes iterative testing and refinement based on user feedback.

- **Collaboration and interaction:** HCAI focuses on fostering effective collaboration between humans and AI systems. It encourages designs that facilitate smooth interaction, where AI systems act as cooperative partners rather than autonomous agents.

2.4 SUMMARY

This chapter has focused on the ethical considerations and governance frameworks essential for the responsible development and deployment of AI systems. It has underscored the importance of a human-centric ethical purpose, ensuring that AI aligns with human values, rights and societal wellbeing. AI ethics, which revolve around transparency, accountability and fairness, are vital for preventing biases, privacy violations and unintended harm.

The chapter has introduced key ethical questions related to AI, such as responsibility, human rights and the challenges posed by advanced AI systems, for example AGI. It has highlighted the importance of ethical frameworks and governance, referencing guidelines from the EU and organisations such as the ISO. The EU's AI Act has been discussed as a pioneering regulatory framework aimed at ensuring that AI respects fundamental rights and promotes transparency and safety.

The chapter has also explored AI governance, which involves setting standards for data usage, algorithmic fairness and decision-making to build trustworthy AI systems. Additionally, the chapter reviewed efforts by various countries, including the US, Japan, Canada and the UK, in setting ethical AI standards, stressing that ethical AI must be part of a broader regulatory and risk management strategy to prevent potential harms.

Finally, ethical concerns such as bias, transparency and the societal impact of AI have been explored, emphasising that a human-centred approach is necessary for developing AI systems that enhance human capabilities and align with societal values. The chapter concluded by discussing risk management frameworks, such as SWOT and PESTLE analyses, and the role of regulation in balancing innovation with the need for safety and fairness.

3 AI ENABLERS, TECHNOLOGIES AND APPLICATIONS

This chapter brings ethical human-centric AI that is robust and trustworthy into something that we can relate to. To do this, we must make sense of how intelligent entities relate to robotics and where the popular term machine learning can easily confuse things. ML is a term that is often used instead of AI. AI and ML are used interchangeably without a clear understanding that ML is a subject that underpins AI. Robotics has been a fundamental driver of AI, learning from experience, and ML, learning from data. Section 3.3 explores how AI relates to AI robots and ML.

3.1 COMMON EXAMPLES OF AI

AI has become an integral part of modern life, manifesting in various forms and applications that touch almost every aspect of our daily routines. Here we discuss some common examples of AI that showcase its versatility and widespread use.

3.1.1 Virtual assistants

Virtual assistants such as Apple's Siri, Google Assistant, Amazon's Alexa and Microsoft's Cortana are prominent examples of AI in action.

These AI systems utilise NLP to understand and respond to user queries, manage schedules, set reminders, control smart home devices and even engage in casual conversation. Their ability to learn from user interactions and improve over time exemplifies ML capabilities. We must be aware that ML answers can be creative or even made up (generated best guess) and lack rigour and accuracy.

3.1.2 Recommendation systems

AI-driven recommendation systems are behind the personalised suggestions we encounter on platforms such as Netflix, Amazon and Spotify. These systems analyse user behaviour, preferences and interactions to suggest movies, products and music that are likely to interest the user. By leveraging techniques such as collaborative filtering and content-based filtering, these AI models enhance user experience and engagement. These are typical examples of learning from data, and use ML.

3.1.3 Fraud detection

In the financial sector, AI plays a crucial role in detecting fraudulent activities. Banks and financial institutions employ ML algorithms to analyse transaction patterns, identify anomalies and flag potentially fraudulent transactions. These systems are designed to adapt and improve continuously, making them increasingly effective at spotting and preventing fraud.

3.1.4 Healthcare

AI is revolutionising healthcare through applications such as diagnostic tools, personalised treatment plans and predictive analytics. AI systems can analyse medical images to detect conditions such as cancer, interpret genetic information to suggest personalised treatments and predict patient outcomes based on historical data. Indeed, some scientific disciplines cannot practise without powerful computers. Here, scientists and engineers use machine learning daily, but they don't necessarily call it AI or ML. Numerical techniques used by engineers and scientists tend to be more effective for their work; however, they have the transferable skills to use ML and AI in developing better work.

3.1.5 Customer service

Many businesses are adopting AI-powered chatbots and virtual customer service agents to handle routine inquiries, provide support and enhance customer experience. These AI systems can process and respond to customer questions, resolve issues and even handle transactions. Their ability to operate 24/7 and learn from interactions helps businesses improve efficiency and customer satisfaction.

3.1.6 Language translation

AI has significantly advanced language translation services, with tools such as Google Translate utilising neural machine translation (NMT) to provide more accurate and natural translations. These systems can translate text and speech in real time, breaking down language barriers and facilitating global communication.

3.1.7 Social media monitoring

AI is extensively used in monitoring and managing content on social media platforms. Algorithms are designed to detect and filter out inappropriate content, spam and fake news. Additionally, AI helps in analysing user engagement, understanding trends and personalising content feeds to keep users engaged.

3.1.8 Smart home devices

AI is embedded in various smart home devices, including thermostats, security systems and household appliances. These devices learn user preferences and behaviours to optimise home environments, enhance security and improve accessibility, independence and convenience.

3.1.9 Robotics

In manufacturing and service industries, AI-driven robots are used for tasks such as assembly line work, quality control and even customer interaction. These robots employ ML to perform complex tasks, adapt to new situations and collaborate with human workers.

These examples of AI's presence are commonplace now, with applications that enhance accessibility, convenience, efficiency and personalisation across various domains. As AI technology continues to advance, its impact on society is likely to expand, leading to even more innovative and transformative applications. The AI agent allows us to relate to AI as a technology or service that helps humans.

3.1.10 Education

AI is transforming education by enhancing both teaching and learning experiences through personalised learning, intelligent tutoring systems and administrative automation. Personalised learning platforms use AI to tailor educational content to individual student needs, learning styles and pace, thereby improving engagement and understanding. Intelligent tutoring systems, such as those developed by Carnegie Learning, provide real-time feedback and support, simulating one-on-one instruction. AI-powered tools facilitate the creation of immersive learning experiences through virtual and augmented reality, making education more interactive and accessible. By leveraging data and advanced algorithms, AI helps to create more efficient, inclusive and effective educational environments.

3.2 AI, MACHINE LEARNING AND ROBOTICS

In this section, we explore AI in terms of agents and robotics. While doing so, we should remind ourselves of the definition of human intelligence discussed in Section 1.1.1.1, from the *Encyclopaedia Britannica* (Sternburg, 2017):

> Human intelligence: mental quality that consists of the abilities to learn from experience, adapt to new situations, understand and handle abstract concepts, and use knowledge to manipulate one's environment.

It is really helpful to use the AI agent as the schematic of the artificial human, and it has a simple analogy in that it is similar to a robot. Unfortunately, this simple analogy has not been used by roboticists. Their focus hasn't been human intelligence or the framework for the academic study of intelligence; instead, they have been focused on the engineering of a robot, which is complex and sophisticated engineering. We are now seeing these disciplines collaborating and exciting new areas emerging. It's an exciting time because advances in ML have unlocked the ability of computers to teach themselves how to achieve a goal, in either a digital simulated environment or a practical engineered environment. It opens up a world of opportunities: autonomous smart cities, smart roads, smart homes, smart TVs, smartphones, smart watches, smart nano-bots.

We can think of AI products, big or small, as AI agents. They can have a perception of the environment they are in and then act on the environment to achieve a goal. ML gives us the digital examples of how this can be achieved. It can also be used to give the AI agent or products autonomy to learn from experience, for example how to undertake a task or tasks and achieve a goal. The functionality we require for intelligent agents links simply with ML. Games can be used to understand the interaction of autonomous agents.

Agent-based modelling is a fundamental method useful in academia, engineering, business, medicine and elsewhere. Through this, we can understand the emergent behaviour of multiple agents operating in an environment.

3.2.1 The AI agent

We start with the academic description of an AI agent from Russell and Norvig (2016). They define a rational agent as:

> For each possible percept sequence, a rational agent should select an action that is expected to maximize its performance measure, given the evidence provided by the percept sequence and whatever built in knowledge the agent has.

This can be illustrated as a learning agent (see Figure 3.1). The AI agent is the schematic of how AI describes a human acting in an environment; the traditional engineering concept of a robot or control system is the dashed line. ML has certainly been used inside this dashed line for many years – we will introduce perceptrons in Chapter 4. With the advent of deep learning and its more recent successes, this dashed line is now moving to encapsulate the whole AI agent. Learning from experience now means the potential for autonomy. We humans sometimes call this 'agency'.

> Agency comes from late middle English (someone or something that produces an effect), founded on the Latin word *agere*, 'to do/act'.

Figure 3.1 The learning agent – based on the structure of agents in Russell and Norvig (2016)

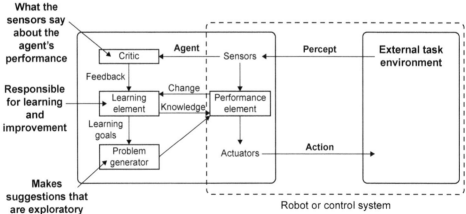

The 'learning agent' is an overarching schematic of an AI agent that can learn how to do something. This specialised agent is capable of learning from its experiences, unlike the four intelligent agents detailed above that act on information provided. Typically, it starts with some basic knowledge and adapts through learning. A learning agent is able to:

- perform tasks;
- analyse its own performance;
- look for new ways to improve on those tasks.

Moreover, it can do this all on its own.

An example of a learning agent is a personal assistant such as Siri or Alexa, which can look up your local weather conditions on the internet and advise you whether to wear a coat or not because it has learned that you like to go for a walk when you haven't slept well.

Humans revere agency, and humans are agents that act in an environment (Coole, 2017). Humans want the free will to act – it is important to us. For a robot to have free will to act, we call this autonomy. Autonomous robots are usually portrayed as baddies in movies because we try to wrap meaning around a robot's actions. Robots, at least for now, are objective and rational and do what we tell them to do (sometimes this may have bad consequences). Humans are subjective, our free will is subjective and when we think about agency we must consider the subjective side of our actions – this leads to the philosophical and moral understanding of humans. In defining an AI agent, we rely on the agent being rational, in the same way that engineers use objective science to design robots.

3.2.1.1 The four rational agent dependencies
We expect AI agents to be rational, and this depends on four things (Russell and Norvig, 2016):

1. The performance measure that defines the criterion of success.
2. The agent's prior knowledge of the environment.
3. The actions that the agent can perform.
4. The agent's percept sequence to date.

These four elements are required to provide a fundamental academic basis for a rational agent. If we think about it, we need all four so that others can reproduce our results. Without them the agent would be ill-defined.

We know that AI is an academic subject based on the scientific method. Rationality helps when we are thinking about AI agents. So, in designing an AI agent – maybe a robot or a data robot or a simulated agent looking at how humans behave – we expect them to act rationally and do what we ask them to do. Going back to Russell and Norvig's definition of the rational agent (2016):

> For each possible percept sequence, a rational agent should select an action that is expected to maximize its performance measure, given the evidence provided by the percept sequence and whatever built in knowledge the agent has.

We can see that when we design an AI system, a robot or an engineered product, it is telling us that it must perform as we design it to perform.

3.2.1.2 Agents – performance measure, environment, actuators and sensors

If we are an engineer asked to design an intelligent entity, then it makes a lot of sense to define a performance measure by which its performance can be measured, define the environment in which it operates and the actuators and sensors it uses to manipulate its environment. By doing so, we create an intelligent entity that can learn from its experience. If we take a humanoid robot builder as an example, then we could describe each as the following:

- Performance measure is the speed to lay a fixed number of bricks in a wall.
- Environment is the building area or room.
- Sensors are tactile sensors, power sensors, visual and sound sensors and so on.
- Actuators are hands, body, feet, legs and so on.

This is a factored state.

Somehow, this seems easy, but it is not explicit mathematically. We could think about building a digital twin so we can model it and learn. By this we mean we can build a simulation environment for the robot to learn from, a bit like a computer game. Just as we teach pilots how to fly in a flight simulator, we can teach a robot in the same way. Or, perhaps, we can let the robot use reinforcement learning to teach itself. For a more detailed understanding of these concepts mathematically, see Russell and Norvig (2016).

An agent might want its own map of the environment. We call this the state of the world of the agent. It is the agent's internal world that it might use to make up for a lack of information. This world can be simple, for example a length of railway track and the temperature of the track might be all we need to control our autonomous train. This is the atomic state. It might be a little more involved where we use a range of sensors to determine the state of the world or environment in which we are operating – for example, rain, temperature, wind and humidity determine the weather that our train is travelling in. This is a factored state. Finally, we have the most complete understanding: the structured state. Here, the objects and the relationship of those objects give the agent a more detailed understanding of the environment it is operating in. It might use AI to predict the outcome of different actions; this is shown schematically in Figure 3.2.

In understanding the agent, we need to understand the mathematics of its internal world, and so the use of a digital twin seems a sensible idea if we need to train an intelligent entity. What happens if we need multiple intelligent agents working together to achieve a goal?

Figure 3.2 The state of the world – the agent's internal world that can make up for incomplete information or help to make decisions

How can we represent the world that supports the making of models?

Increasing fidelity

Atomic state: a black box with no internal structure.

Factored state: a vector (list) of attributes made up of Booleans, real-valued or one of a fixed set of states.

Structured state: made up of objects (could have its own attributes) as well as relationships with other objects.

We humans describe most things using objects and relationships via natural language – engineers and scientists use these descriptions to build products, services and do research

3.2.1.3 Agent-based modelling

Agent-based modelling (ABM) is used to understand how agents working together can achieve a goal (Wikipedia Contributors, 2024l). We might think of this as how bees build a wonderful structure to produce honey. An agent-based model can look at how the bees communicate, how they maintain order or control; also, how an intelligent entity would work around humans or other animals. ABM is a really useful approach to understanding emergent behaviour. When we think about ABM, controlling multiple agents or finding a hierarchical controls system like subsumption will become a key concept (Wikipedia Contributors, 2024m). This is a growing area of interest, and we touch on ABM later when we look at intelligent robotics.

3.2.2 Types of agent: reflex, model-based reflex, goal-based and utility-based

In AI, an intelligent agent refers to an autonomous entity that acts – directing its activity towards achieving goals in an environment using observation through sensors and consequently using actuators – and is therefore intelligent. There are basically four rational types of agent, examples ranging from a relatively simple heating system control agent, building up in complexity to an automated medical general practitioner (GP/doctor) agent. Let's look at each of them.

3.2.2.1 Reflex agent

In AI, a simple reflex agent is a type of intelligent agent that performs actions based solely on the current situation. The agent does this through predetermined rules for these conditions. This is commonly referred to as the condition–action rule: the program selects actions based on the current percept. It's simple to understand and program.

Example: the central heating overheats, so the program selects the action to switch off the power.

39

3.2.2.2 Model-based reflex agent

In a model-based reflex agent, the agent has a model of the world that it can call upon if needed:

- It can make up for a lack of sensors by using a virtual world that doesn't need sensor data.
- The program now looks at the percept and updates its own internal world (state).
- The program can then assess the possible actions and future states, and uses the reflex agent approach to determine what actions to take.

Example: an underwater vehicle loses visual sensor data from stirring up silt – there's not enough light – but it can continue on using its 3D mapped geometrical model.

3.2.2.3 Goal-based agent

Goal-based agents can further expand on the capabilities of the model-based reflex agents by using 'goal' information. The goal information describes situations that are desirable.

Search and planning are the subfields of AI devoted to finding action sequences that achieve the agent's goals:

- The program needs more than just sensors and an internal world to implement the agent's functionality: it needs a goal.
- These programs are more versatile and flexible and can adapt to changes.

Example: an autonomous vehicle can choose one of five exits from a motorway – goals could be the safest, quickest, most scenic, cheapest and/or shortest route.

3.2.2.4 Utility-based reflex agent

A utility-based agent is an agent that is a step above the goal-based agent. Its actions are based not only on what the goal is, but the best way to reach that goal. In short, it is the usefulness or utility of the agent that makes it distinct from its counterparts. The utility-based agent makes its decisions based on which action is the most pleasing or the most effective at achieving a goal. The goal is measured in terms of the agent's utility.

Utility is the scientific way economists measure an agent's happiness; it is more versatile than a simple binary yes/no or happy/unhappy output, it actually measures how useful it is.

Example: an AI chatbot doctor assesses the efficacy of a treatment based on a patient's individual medical history and current health needs – that is, what utility is this medication to the patient?

3.2.2.5 The relationship of AI agents with machine learning

An agent should be autonomous. What this means for an intelligent entity is that it can learn, and make up for incomplete knowledge and incorrect prior knowledge. For example, the Deep Blue machine that beat the world chess champion had not learned how to do that by itself. It relied on the prior knowledge of the programmers to win its games.

A truly rational agent should be autonomous – it should learn to make up for incomplete and incorrect prior knowledge – but we think it's safe to say that we are concerned about agents having too much autonomy. It should be recognised that, at present, a robot can't just wake up one day and start doing something; we need to be realistic and give the robot a sensible starting point.

An agent lacks autonomy if it relies on prior knowledge from its designer, but there needs to be some prior knowledge to get it going. It will certainly need an initial condition.

3.2.3 Role of robotics in AI

As discussed in Chapter 1, the word 'robot' came into use in Czech-born Karel Čapek's 1920 stage play *R.U.R.*, which stands for *Rossumovi Univerzální Roboti* (Rossum's Universal Robots). The play opens in a factory that makes artificial people called 'robots'. The robots in the play are closer to the modern idea of androids or even clones, as they can be mistaken for humans and can think for themselves. Without giving the plot away too much, after initially being happy working for humans, a hostile robot rebellion then ensues that leads to the extinction of the human race. As we mentioned earlier, the word 'robot' comes from the Slavic language word *robota*, meaning 'forced labourer'. Nowadays we consider a robot to be a machine that can carry out a complex series of tasks automatically, either with or without intelligence.

Robots have been around for more than 60 years. In 1954 George Devol invented the first digitally operated and programmable robot, called the Unimate. In 1956 he and his partner Joseph Engelberger formed the world's first robot company, and the Unimate, the first industrial robot, went online in the General Motors automobile factory in New Jersey, USA, in 1961. Robot development has been ongoing around the world ever since, and robots in various guises are now in general use in most factories and distribution centres around the world, quite often working with and alongside humans on complex production lines.

You may even have allowed domestic robots to invade your own home – robotic vacuum cleaners, first appearing in 1996, are the most widely recognised. The Roomba, first introduced in 2002, is an autonomous robotic floor vacuum cleaner with intelligent programming. Despite nearly 20 years of development, AI home helpers still struggle with obstacles such as dog faeces, cables and shoes. Similar technologies are often employed to mow lawns, but also struggle with discarded garden toys and, of course, humans' best friend.

One of the most widely recognised robots is ASIMO (Advanced Step in Innovative Mobility), which is a small child-like humanoid robot created by Honda in 2000 (see Figure 3.3). It has been developed continually since then and now has become one of the world's most advanced social robots. Unfortunately, ASIMO has recently gone into retirement, but Honda say that the technology will be diverted into 'nursing care' robots in the future.

Figure 3.3 Honda's ASIMO, conducting pose captured on 14 April 2008 (https://commons.
wikimedia.org/wiki/File:ASIMO_Conducting_Pose_on_4.14.2008.jpg; Creative Commons Attribution-Share
Alike 3.0 Unported licence)

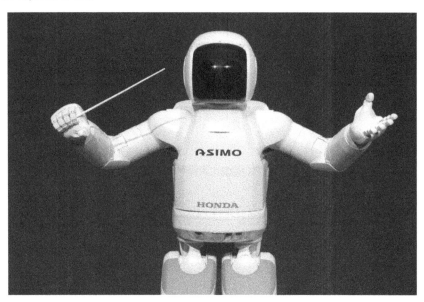

Although many people look towards the positive aspects of robot development,
others see robots in a negative light and project a future in which robots take jobs
from human workers. However, 'crystal ball gazers' generally believe that the AI and
robotics industries will probably create as many new jobs as are lost through increased
automation.

Although we have focused so far on intelligent robots, the vast majority in operation are
unintelligent (i.e. there is no learning) and simply follow a computer program to carry
out a limited action. Now, if we apply AI techniques to robots and allow them to learn and
work autonomously across a number of areas, we could equally have created either a
useful home assistive robot or the robot from *The Terminator*, depending on your point
of view.

3.2.3.1 Robotic paradigms
A paradigm is a philosophy or set of assumptions and/or techniques that characterise
an approach to a class of problems. The word paradigm comes up a lot in the academic,
scientific and business worlds. When you change paradigms, you are changing how you
think about something; it is a way of looking at the world – it is also a set of tools for
handling and solving robotic problems.

Specifically, in robotics, a robotic paradigm is a mental model of how a robot operates.
A robotic paradigm can be described by the relationship between the three primitives
of robotics: sense, plan and act. It can also be described by how sensory data are
processed and distributed through the system, and where decisions are made.

In robotics, there are three paradigms:

1. hierarchical;
2. reactive;
3. hybrid hierarchical–reactive.

These three paradigms are used for organising intelligence in robots – and selecting the right paradigm makes problem-solving easier. The hierarchical, reactive and hybrid hierarchical–reactive paradigms all use two or three robotic primitives:

- sense (uses information from sensors and the output is sensed information);
- plan (sensed or cognitive information and the output is a directive);
- act (sensed information or directives and the outputs are actuator commands).

In robotics there are two ways to describe a paradigm: either by the relationship of sense, plan and act, functions the robot undertakes are categorised into the three primitives; or by the way sensor data are utilised and organised:

- How is the robot or system influenced by what it senses?
- Does the robot take in all sensor data and process them; for example, or are sensor data processed locally?

3.2.3.2 Hierarchical
Here, there is a fixed order of events, which is based on top-down planning (see Figure 3.4). These robots tend to have a central model to explicitly plan from. It tends to be highly controlling in a highly controlled environment and difficult to use in complex environments. This is probably OK in a highly organised and engineered world, but it does ignore the human cognitive and biological elements; for example, we can act and sense at the same time. An example is a robot that selects an item from a fixed location.

Figure 3.4 Hierarchical paradigm in robotic design

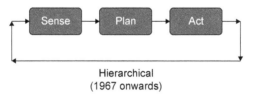

Hierarchical
(1967 onwards)

3.2.3.3 Reactive
Reactive robots are always sensing and acting. As the sensing changes, so its behaviour changes (see Figure 3.5). Couple these behaviours together and you are not stuck with one sense action, which saves planning time. To gain a more intuitive understanding, it's like the knee jerk reaction to being surprised – if we place our hand on a hot surface we pull away really quickly, and that is automatic. With the reactive paradigm, the robot reacts to a sense input. An example is a robot that maintains a set direction but, if a sensor senses an obstacle, the wheels react and turn away from the obstacle.

Figure 3.5 Reactive paradigm in robotic design

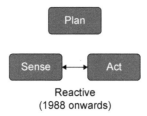

3.2.3.4 Hybrid
This paradigm is a combination of the hierarchical and reactive paradigms. These robots plan their initial action to achieve a goal. Sense and act are then informed by a plan (see Figure 3.6).

Figure 3.6 Hybrid deliberative/reactive paradigm in robotic design

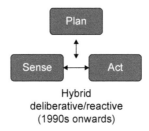

Sensors can update the plan (e.g. identify an object), and this ability allows each of the sensors, actions and planning to run at a speed that is appropriate. This paradigm is a natural evolution of robotic design that allows different parts of the robot to be controlled by something that is appropriate for what it is doing. Some of our cars have automatic braking systems that help us to brake more effectively if there is an obstacle or an icy road. These systems are reactive, but we can think of the driver as the hybrid deliberative/reactive part of the system – the driver does the sense, plan, act. An engine management system can be thought of as a 'sense – plan, act' robotic paradigm: it takes multiple readings from the engine (temperature, engine load, air flow rate), plans the optimum fuel to air mixture and ensures the car delivers the maximum power with the least emissions.

3.2.3.5 What is an intelligent robot?
This is possibly the shortest section on our topic of robots. An intelligent robot is one that uses AI or, as Robin Murphy defines it (Murphy, 2019), 'an intelligent robot is a mechanical creature which can function autonomously'.

We must note that the idea of a robot here is a mechanical robot that we see in industry or a humanoid robot that could be our personal assistant. Robots can also be found in the data world of digital computers. Here, robotic process automation (RPA) is a term used to describe agents that work with data in an enterprise setting. RPA isn't usually an intelligent agent, even though the rules it follows are complex.

3.2.3.6 Intelligent robotics to intelligent agents

The advent of the intelligent learning agent plus reinforcement learning NNs making significant progress over the past few decades have laid the foundation for the concept of autonomous robots to become a reality. Autonomous vehicles are commonplace throughout the United States now; many states allow testing of robotic vehicles in a driverless setting. Ideas such as ABM, digital twins and simulation are maturing into fertile ground for development – which really means learning from experience. Complete ecosystems for robotic development are closely linked with learning environments for AI agents. ABM involving humans and machines seems the next natural step – multiple agents, multiple intelligent robots and, of course, multiple intelligent humans interacting.

3.2.4 Machine learning basics

ML is about learning from data. What we see today, in 2024, is the rapid growth and adoption of narrow ML. We have concentrated on narrow AI, and one of the reasons is that digital computers are limited by the processing power, energy and accuracy of today's technology. As a result, we can only concentrate on narrow ML, focused on specific well-defined tasks or goals. These tasks are defined by Tom Mitchell's often quoted definition (Mitchell, 2018: 2):

> A computer program is said to learn from experience, E, with respect to some class of tasks, T, and performance measure, P, if its performance at tasks in, T, as measured by, P, improves with experience, E.

The examples given of these types of tasks are playing games such as chess, checkers and draughts. Modern-day games include simulation games and very advanced strategic, well-engineered games such as Go; these types of games can be explicitly defined on a digital computer. Practical examples in the real world include optimising where aircraft park at an airport or the logistics of delivering a parcel; again, a reasonably well-ordered and engineered environment in which ML can optimise something.

ML is focused on explicitly defining a problem that can be solved on a computer. These problems can be complicated, non-linear and statistical. In simple terms, if we are to use ML in our AI, we must be able to represent our problem mathematically and in such a way that it can be solved by a machine. Today, we typically use digital computers. However, quantum, analogue, optical and biological computers are on their way.

Digital ML has become very popular recently with the success of convolutional deep NNs. These numerical techniques have given us an understanding of how the human mind solves problems. So, when we think about ML, we can think about an AI agent learning from data. These machines are now so good at playing games that they can beat the world champions.

The AI agent is much more than a narrowly focused digital computer program. ML works on data in the computer. We must work really hard to think of this as an interface with actuators and sensors in an environment. It is sensible to think, here, that ML learns from data in a computational environment. This is a good starting point to opening up the world of AI. AI is about humans and machines working together to achieve goals. We might even go on to say that ML is an AI enabler setting the foundation for a future of humans and machines.

3.2.4.1 How we learn from data

ML is widely available nowadays and it is an enabler for more involved AI as we accelerate our applications and build intelligent entities. ML is built by defining what functionality we need, writing software to implement that functionality and then constructing the machines (typically digital machines) to obtain our functionality. So, our intuitive and simple model is functionality, software and hardware. AI is about learning from experience and, if we have lots of data, what type of functionality is needed to learn from those data.

3.2.4.2 Introductory theory of machine learning

At the core of AI is ML. In fact, ML is an enabler of AI. ML is most often associated with the popular computation we have today, that is, digital computation. AI is about learning from experience and ML is about learning from data. We will use this as the basis of our intuitive understanding of what ML is. The most popular machines used today are digital personal computers, computer clusters and large high-performance computers. As we look forward, we will move towards optical, quantum, biological and other types of learning machines to help us. And now, with our smart devices with onboard AI hardware, we see a growth in edge computing where digital ML is done on our smart devices such as phones, speakers and cars. Security is a key concept here; how do we ensure that our data are safe?

To start, we need to think about what we are actually doing – we are going to instruct and teach a machine to help us with learning. If we use a digital machine, this means we need to understand our problem in such a way that we can let a machine, say a laptop or smartphone, do the learning for us. The hardware needs software and the software needs an algorithm and data. The algorithm will use mathematics, and we finally have our starting point.

In 2019 Gilbert Strang, a professor at the Massachusetts Institute of Technology, wrote a book on linear algebra and learning from data (Strang, 2019). He tells us that ML has three pillars: linear algebra; probability and statistics; and optimisation. Much earlier, in the 1990s, Graham, Knuth and Patashnik produced a textbook to help students understand the mathematics needed for computational science – we might think of it as the book that describes algorithms mathematically, how they are defined and how they are understood (Graham et al., 1994). These two publications are detailed, rigorous mathematical texts. Tom Mitchell's book on machine learning, as mentioned in Chapter 1, gives a taste of the theory of ML with examples, and is again a detailed, rigorous mathematical text (Mitchell, 2018). These books are comprehensive and not for the novice. They are relevant to mathematicians, engineers and physicists who are well versed in these types of disciplines.

Strang's *Linear Algebra and Learning From Data* (2019) goes into the theory in depth and refers to many excellent examples and references for scientists, engineers and mathematicians.

Graham, Knuth and Patashnik's *Concrete Mathematics* (1994) is a book about the mathematics for computer science. It is more detailed in its explanations of algorithms and theory, and introduces statistics and probability. When coupled with Knuth's *The Art of Computer Programming* (2011), these books provide an extensive reference literature on the nature of learning using computers, digital and other resources.

AI is more than learning from data; when we think about AI products, we must also think about the world that they operate in, how actuators are controlled and how policies are decided given sensor readings and a percept. Stuart Russell and Peter Norvig's book on AI goes into the theory and mathematics of how we frame our problem beyond just learning from data (Russell and Norvig, 2016); they cover ML too. They are also very clear that AI is a universal subject, so can be applied to any intellectual task – learning from experience.

What is common to these technical books on the theory of AI are some key subject areas that are worth gaining an appreciation of. These are: linear algebra; vector calculus; and probability and statistics. Linear algebra is about linear systems of equations; vector calculus is about differentiation and integration of vector spaces, or, more practically, how mathematicians, engineers and physicists describe our physical world in terms of differences and summations and probability; and statistics is about the mathematics of understanding randomness.

3.2.4.3 Linear algebra
This section is designed to give an overview of linear algebra; as Gilbert Strang explains, linear algebra is the key subject to understanding data and is especially important to engineering (Strang, 2016).

We learn from data using linear algebra. In school most of us learn about systems of equations or simultaneous equations. It is linear algebra that we use to solve equations using computers, and it is at the heart of AI, ML, engineering simulation and just about every other subject that relies on digital computers.

A simple example will introduce what we mean. If we want a computer to learn, we have to represent that learning in the form of mathematical operations that a computer can undertake. It is the terrain or geometry that computers can work with. Computers can undertake these mathematical operations, such as add, subtract, divide and multiply, very quickly and accurately. Far quicker and more accurately than us humans. As we mentioned earlier, most of the problems in the real world require an adaptive or iterative approach – and computers are good at this too. In fact, they are so good at this, and quick, we might think it is easy. In doing so, we almost trivialise the amount of work that goes into preparing our machines for learning.

The simplest example of how we do this is a problem we learn to do at school, solving simultaneous equations. This can be done by hand using a technique from the 1800s: Gaussian elimination (Chinese mathematicians were aware of this technique before Gauss: Wikipedia Contributors, 2024n). We could do this by hand with a calculator, a set of log tables or an analogue computer – the slide rule – however, digital computers can do this quicker and more accurately than we can. If our problem has more unknowns, say five or even 1,000, then this would be impossible to do by hand.

To solve this problem with a computer we need to write it in a slightly different way, and to do this we can use linear algebra. Scientists, engineers and mathematicians use these techniques to set up problems with large matrices, and have to invert big matrices. This requires high-performance parallel computers, but it can be done. AI draws on these techniques also, and is another user of large supercomputers.

If we are to formulate a learning problem that we want a digital computer to solve, then we need to understand linear algebra. There are already multi-billion-dollar industries (finite element analysis, computational fluid dynamics, business analytics, etc.) using these techniques.

Armed with machines that we can use to enhance our learning, we now move onto the next core topic from mathematics, that of vector calculus.

3.2.4.4 Vector calculus

Vector calculus allows us to understand vector fields using differences and summation. Or, to use the mathematical terms, differentiation and integration. A good example of a vector field is the transfer of heat. Heat can move in three directions, and we observe that it moves from hotter sources to cooler sources. So, in a simple model, we can think that the amount of heat that moves between two bodies depends on the difference in temperature. If we monitor this over a long period of time, we can add these measurements up and calculate the total amount of heat that has been transferred. This is vector calculus; it allows us to represent problems that we encounter day-to-day, from the Earth's electromagnetic field, the movement of the oceans to the motion of vehicles on a road. When we think of a learning agent, we need to represent the state of the world. When we use the analogy of a learning problem as a terrain, then vector calculus will help us to understand how our AI is learning. In particular, how quickly it learns.

Mathematicians, scientists and engineers all work with partial and ordinary differential equations to solve their problems. Vector calculus is the cornerstone building block. If they are really lucky, there is an analytical solution to the problem; more often than not, this is not the case and they revert to numerical solutions. We are guided back towards using linear algebra. The linear algebra is used by digital computers to solve the numerical problems.

Dorothy Vaughan used these techniques in the 1950s to help NASA land a vehicle on the moon and return it safely (The Editors of Encyclopaedia Britannica, 2024a). More generally, we don't need to think about vector calculus applying only to space, that is, Euclidean space of x, y and z directions. The ideas and concepts can be generalised to any set of variables, and this is called multi-variable calculus. We can effectively map out a space made up of different variables and use differentiation and integration to understand what is happening.

If we were training a NN, we could map out the training time in terms of number of layers, number of nodes, bias, breadth and width of the layers, then understand how effective our training is using vector calculus. For example, we might ask what gives us the minimum error in the quickest time.

Models of the state of the world (agent's internal worlds) or ideas such as digital twinning use these techniques. These ideas work quite well on engineered systems where we have some certainty on the system we are working with. The real world, however, is not perfect and in AI we also need to deal with uncertainty or randomness. This is the next mathematical topic that AI needs – probability and statistics.

3.2.4.5 Probability and statistics

Probability and statistics are vital because the world we live in is random. The weather, traffic, stock markets, earthquakes and chance interfere with every well thought out plan. In the previous sections we introduced linear algebra and vector calculus, which are elegant subjects but also complex and need the careful attention of a domain expert. Randomness adds another dimension to the types of problem we might face with AI.

Applied mathematicians, engineers and physicists all use statistics to extend the capabilities of linear algebra and vector calculus to understand the world we live in. This is easier to observe in the diagram at Figure 3.7. Some examples are helpful here. The low-randomness, low-complexity learning is what we can work out as humans, perhaps with a calculator. For example, timetables – how long it takes for a train to travel from London to Kent. We can make these calculations easily. The low-randomness, high-complexity quadrant is learning that involves complex ideas or theory, like linear algebra or vector calculus. An example of this is how Dorothy Vaughan used computers and Fortran to calculate the re-entry of a spacecraft into Earth's atmosphere. Low-complexity and high-randomness learning is where we can use simple ideas to understand randomness. Examples of this are games of chance, such as card games, flipping a coin or rolling dice. Physical examples are the shedding of vortices in a river as it passes around the leg of a bridge. High complexity and high randomness is learning that involves complex ideas that also include high randomness; examples are the weather, illness in patients, quantum mechanics, predicting the stock market. They often involve non-linearity, discontinuity and combinatorial explosions.

Dorothy Vaughan's work using computers to calculate the re-entry point of a spacecraft into the Earth's atmosphere was an elegant example of using a computer to iterate to find a solution – something that we use a lot in learning. Learning is non-linear and iterative. Dorothy took a complex problem and simplified it so a computer could guess repeatedly and find a solution. We can think of this as a heuristic that allowed a digital machine, using Fortran, to guess a solution. This is shown in Figure 3.8.

Statistical analysis of large volumes of data is a modern-day success of digital computers. Without them we would not be able to analyse the information encoded in our genetics. Business analysts are making our activities more effective and efficient, leading to better supply chains that are integrated. An intuitive understanding of statistics and probability will help those working in AI. In AI, inference and the ability to

Figure 3.7 Complexity and randomness

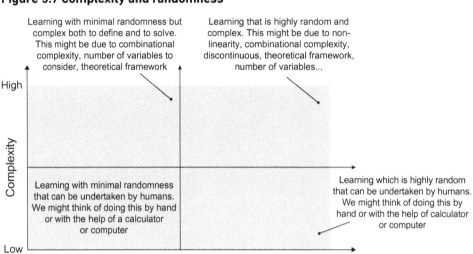

Figure 3.8 Heuristics can simplify complexity and randomness

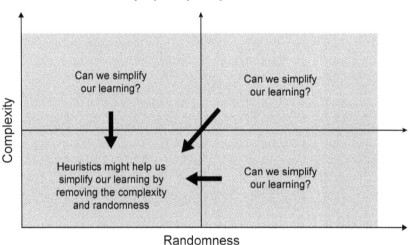

infer something about data are of particular importance. This started some time ago with Reverend Thomas Bayes. He came up with a theory (Bayes' theorem) that allowed us to infer the probability or likelihood of an event happening given knowledge of factors that contribute to the event (Routledge, 2005). We need to remind ourselves that the factors may or may not cause the event, but we have noticed that, if they occur, it could also mean that the event will happen. For example, it rains when there are clouds in the sky (50 per cent of the time), the temperature is 19°C (33 per cent of the time), the pressure is ambient (22 per cent of the time) and so on. We can now ask ourselves what would be the probability of it raining if there are no clouds in the sky and the temperature is 19°C? Can we infer this from the data we have already?

To understand Bayes' theorem we need to go back to basics and start with Russian mathematician, Andrey Nikolayevich Kolmogorov (Lee Cooke, 2020). Kolmogorov's contribution to probability and applied mathematics is astonishing. He wrote the three axioms of statistics. To understand them, and the theories that underpin them, we will use Venn diagrams, a pictorial way to represent statistics (Duignan, 2010).

In AI there are two main types of statistical learning: regression and classification (Wikipedia Contributors, 2024o). Classification tries to classify data into labels, for example name a person in a picture. Regression tries to find a relationship between variables, for example the relationship between temperature and heat transfer. This brings us nicely to how we represent random data and how we generate random numbers. Perhaps we have a limited set of statistical data and want more data points? This requires an understanding of how we represent random data.

Random data come in two types: discrete and continuous. An example of discrete random data is the score from the roll of a die. An example of continuous random data is the output from a pressure sensor recording the pressure on the skin of an aircraft's wing. It is useful to know how these random data are distributed.

For discrete data we use a probability mass function, and this is shown in Figure 3.9. It is sometimes called the discrete density function. In this example we have a histogram showing the possible outcomes of rolling two dice. They are discrete values because we can only obtain the following scores from adding the two:

2, 3, 4, 5, 6, 7, 8, 9, 10, 11 and 12.

Figure 3.9 Example of a probability mass function of rolling two dice (Adapted from Tim Stellmach, source: https://commons.wikimedia.org/wiki/File:Dice_Distribution_(bar).svg)

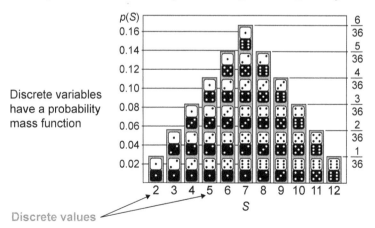

We can see that there is one way to obtain the score 2 and six ways of obtaining the score 7. The probabilities of obtaining each score, 2 to 12, add up to 1.

Continuous random data are represented by probability density functions (PDFs) (The Editors of Encyclopaedia Britannica, 2024b). Probability is now thought of as being in a range, say in between two values. So, from our example of the pressure acting on the skin of an aircraft's wing, what is the probability of the pressure being between 10Pa and 15Pa above atmosphere? Or we can think of the traditional way we grade exam scores: different ranges are associated with the grades received by the people taking the exam. This is shown in Figure 3.10. Here we see a standard Gaussian curve and the mean score is in the middle. Particular grades are associated with the bands in the curve that define a score range.

Figure 3.10 Example of a probability density function of continuous data (https://commons.wikimedia.org/wiki/File:Standard_deviation_diagram.svg under Wikimedia Commons licence)

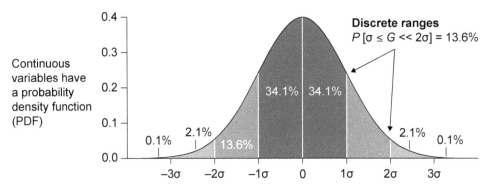

PDFs come in many forms and have different measures to describe them. By describing our discrete or continuous random data via a distribution, we could also generate random numbers, and we will come back to this later. There is another important idea that can be described by example.

If we have five dice, and use them to generate random scores, we could ask what the distributions would look like; in particular, what happens as we increase the number of dice that contribute each set of scores we generate. This experiment is shown graphically in Figure 3.11. Rolling one die, as we expect, is a uniform distribution because each score is equally probable. When we roll two dice, the distribution is no longer uniform. As we increase the number of dice, the distribution converges onto a normal distribution. The theorem that describes this is the central limit theorem (Bronshteïn et al., 2015). The central limit theorem suggests that, under certain (fairly common) conditions, the sum of many random variables will have an approximately normal distribution. So, if an event is a combination of lots of random data with different distributions, it is likely our event's distribution will be Gaussian or normal.

From our intuitive understanding of statistics and probability we can express random data in terms of distributions and how random events relate to each other. However, we are missing how we generate random numbers or random data. Computers are explicit, linear, precise and deterministic.

Figure 3.11 An example of the central limit theorem using five dice and rolling set of scores. Top left is a set of scores from rolling one die ($n = 1$); middle right is a set of scores from rolling five dice ($n = 5$) (https://commons.wikimedia.org/wiki/File:Dice_sum_central_limit_theorem.svg under Wikimedia Commons licence)

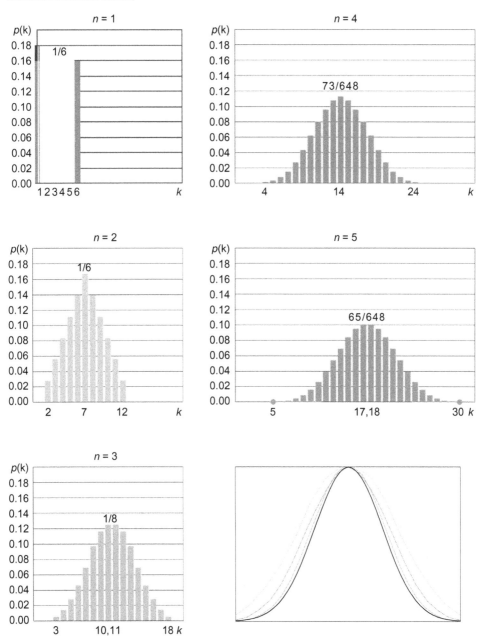

Fortunately, they can produce pseudo-random numbers, and the algorithms to do this are already written (Teukolsky et al., 1993; Knuth, 2011). If we have incomplete data, but can somehow determine the probability distribution from the data we have, then we can generate random numbers with the same distribution.

This section has built up an intuitive understanding of the pillars of ML and AI. Domain experts in these areas are needed if you are working with AI at an algorithmic or training level, so you will more than likely need to have these people in your team. What we have tried to show in this section is the level of mathematical understanding someone implementing AI needs to have when working with AI software. A programmer just learning Python, for example, is not prepared to develop AI algorithms; however, if a programmer is confident in the material in this section, then some higher-level mathematical training will make their job much easier.

3.3 MACHINE LEARNING CONCEPTS

In this section, we describe some common techniques that we use to learn from data. These are high-level descriptions that get us started. There are numerous and abundant examples in the Further reading list and on the internet to learn from.

3.3.1 Common machine learning concepts

Here, we introduce supervised, unsupervised, semi-supervised and reinforcement learning. We describe some techniques for learning from data in Chapter 4, along with how we typically visualise data while stepping through the stages of a ML project.

3.3.2 Supervised, unsupervised, semi-supervised and reinforcement learning

If we have prepared our data well and are confident that the data can be used to test a hypothesis, then we can begin to understand the types of algorithms we employ in ML. What we mean by 'type' is the broad categories that define our algorithms. They are typically organised as follows (Russell and Norvig, 2016):

- Supervised learning – these types of algorithm use labelled (actual solution) data to train the algorithm (see Figure 3.12).

- Unsupervised learning – these types of algorithm use unlabelled data to train the algorithm (see Figure 3.13).

- Semi-supervised learning – these types of algorithm use labelled and unlabelled data to train the algorithm (see Figure 3.14).

- Reinforcement learning – these types of algorithm use the learning of an agent to achieve a goal that is measured in terms of a reward or penalty (see Figure 3.15).

Examples of each of the types of ML are:

- Supervised learning – classification, regression (linear, logistic), SVMs, decision forests, NNs.

- Unsupervised learning – clustering, association rule learning, dimensionality reduction, NNs.

- Semi-supervised learning – combinations of supervised and unsupervised algorithms.

- Reinforcement learning – Monte Carlo, direct policy search, temporal difference learning.

Reinforcement learning fits naturally with the AI learning agent schematic. It relates to both products (e.g. autonomous robots) and data analysis. In reinforcement learning, an agent interacts with an environment to achieve a goal. It can also learn how to achieve that goal.

We can also categorise the way an algorithm learns in terms of the data it is working with.

Batch learning describes learning from large data sets. All of the data are used to train and test the algorithm. The computer resources required are governed by the volume, velocity, variety and veracity of data. This learning is done offline. Online learning is undertaken with data in small or mini batches. Learning occurs as data become available – an example is a system that learns from stock market prices.

There will be times when we come across types of learning from examples or learning from examples to build a model; these types of learning are called instance-based learning and model-based learning, respectively. These are possibly the simplest form of learning.

___

Figure 3.12 Supervised learning

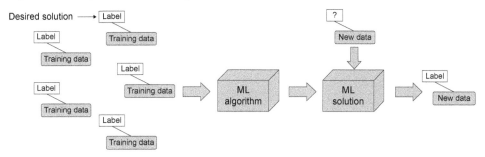

Figure 3.13 Unsupervised learning

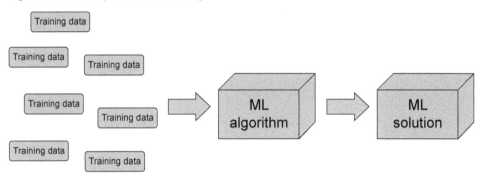

Figure 3.14 Semi-supervised learning

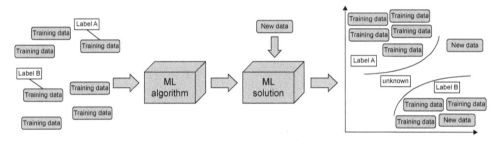

Figure 3.15 Reinforcement learning

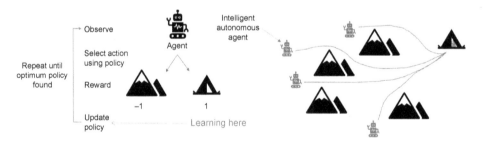

3.4 SUMMARY

This chapter explored how AI, ML and robotics interrelate, emphasising the importance of a human-centric ethical approach. It clarified that, while AI and ML are often used interchangeably, ML is a subset of AI, focusing on learning from data, whereas AI encompasses broader capabilities, including learning from experience and making decisions autonomously. The chapter discussed how intelligent agents, which can perceive and act in environments, are fundamental to both AI and robotics. These agents can learn and adapt, whether in simulated digital environments or physical, engineered ones.

The concept of AI agents was explained through rational agents, which select actions based on maximising performance within a given environment. The chapter detailed various types of agents – reflex, model-based, goal-based and utility-based – highlighting their increasing complexity in handling tasks autonomously.

The role of robotics in AI was explored, tracing the development of intelligent robots and how AI enhances their ability to perform complex tasks. Key paradigms in robotics – hierarchical, reactive and hybrid – were introduced, showcasing how robots process sensory data and make decisions.

The chapter also delved into ML, discussing its foundational principles in linear algebra, vector calculus and probability, as well as the role of supervised, unsupervised, semi-supervised and reinforcement learning techniques in AI applications. These techniques enable AI systems to learn from data, solve complex problems and interact with their environments.

In summary, this chapter connected the technical foundations of AI and ML with practical applications, offering insights into how intelligent agents and robotics contribute to advancing AI technologies in various fields.

4   DATA AND AI

4.1 DATA IN AI

In this chapter, we expand the use of data to learning from it through ML. We may obtain data from various sources, and it will come in many forms and levels of quality. To make sense of this, we will use digital computation as our method for learning from data.

4.1.1 Key data terms

Let's define the typical key terms that we explore in this section. These are:

- data types (structured, semi-structured and unstructured);
- big data;
- data visualisation.

4.1.1.1 Data types – structured data

Structured data in ML refer to highly organised information that is easily searchable and typically stored in tabular formats such as spreadsheets or databases. These data are characterised by consistent, predefined schema with clearly defined rows and columns, where each column represents a specific feature or attribute, and each row represents an individual record or instance. Examples of structured data include customer information databases, financial records and sensor data from IoT devices. In ML, structured data are crucial as they allow for straightforward analysis and the application of various algorithms for tasks such as classification, regression and clustering. Their well-defined nature facilitates data pre-processing, feature extraction and integration with ML pipelines, enabling the development of predictive models that can uncover patterns and insights, leading to data-driven decision-making and automation in numerous applications.

A word of warning: even though this is the simplest form of data to understand and use ML algorithms with, we will need to become proficient in the linear algebra or scalars, arrays, vectors and tensors. In particular, if we take advantage of the vast wealth of open-source ML algorithms available, we must prepare the data and send data to the algorithm in the form it is expecting. It is much more involved than defining arrays and lists in computer code.

4.1.1.2 Data types – semi-structured data

Semi-structured data in ML refer to information that does not conform to a rigid, tabular format but still contains organisational markers to separate semantic elements, making them more flexible than structured data yet more organised than unstructured data. Examples include JSON, XML and YAML files, as well as email messages and HTML documents. This type of data typically includes tags or keys that provide context and hierarchy, allowing for easier parsing and interpretation than unstructured data. In ML, handling semi-structured data involves techniques for extracting and transforming relevant information into a more analysable format. This often requires sophisticated pre-processing steps, such as parsing, normalisation and feature extraction, to prepare the data for modelling. Semi-structured data are valuable because they capture complex, nuanced information from various sources, such as web pages and social media feeds, enabling ML algorithms to leverage richer data sets for improved predictive performance and more insightful analytics.

In essence, the more we practise working with these data the better at extracting value from them we will get. It is more of an orchestration rather than simply following a typical ML project checklist.

4.1.1.3 Data types – unstructured data

Unstructured data in ML refer to information that lacks a predefined format or organisational framework, making it inherently more complex to process and analyse. This data type includes text documents, images, audio files, videos, social media posts and other forms of multimedia content. Due to their free-form nature, unstructured data do not fit neatly into rows and columns, posing significant challenges for traditional data processing and analysis methods. In ML, extracting meaningful insights from unstructured data involves advanced techniques such as NLP for text, computer vision for images and videos and audio signal processing for sound. These techniques enable the transformation of unstructured data into a structured format that ML algorithms can utilise. The ability to harness unstructured data is increasingly important, as it represents a vast majority of the data generated in the digital age, providing a rich source of information for developing predictive models, improving decision-making processes and uncovering hidden patterns and trends across diverse applications.

Working and extracting the value from this type of data requires an expert grasp of working with structured and semi-structured data. Indeed, it may require the use of ML techniques to understand these data. This is an exciting area of ML that we now have at our fingertips. Our smartphones will soon have AI that will better inform our use of them. It may be that our digital AI capabilities will be developed for us, and we won't need to work with unstructured data as much as we think. We'll just become used to having the AI do it for us. What we mean here is that we don't need to understand how the search engine works, and we can maximise its benefit by learning how to use it.

4.1.1.4 Big data

Big data is a term that often describes the vast array of data that can be found on the internet. Big data has the following characteristics (Gregersen, 2024):

- 'Volume measures the quantity of stored and generated data.
- Variety measures the nature and type of data.

- Velocity measures the speed at which the data is moving.
- Veracity measures the data quality, value or utility.'

Big data in ML refers to the massive volumes of data generated at high velocity and with a wide variety of formats that traditional data-processing tools cannot efficiently handle. These data encompass structured, semi-structured and unstructured information originating from numerous sources such as social media, sensors, transactional systems and multimedia. In the context of ML, big data provides the raw material needed to train more accurate and robust models. The sheer scale and diversity of big data enable the discovery of complex patterns, relationships and insights that smaller data sets might miss. Leveraging big data requires specialised technologies and architectures. ML algorithms applied to big data can lead to breakthroughs in predictive analytics, real-time processing and personalised experiences, driving innovation and competitive advantage across industries from healthcare to finance, and beyond.

We might make sense of big data by thinking more about what enables it in the first place. The internet gives us interconnectivity that means data can be dug from all over the world. It also allows us to access the vast amount of data processing we need to work with big data sets. So, to harness big data, we must be familiar with parallel computing and networks.

4.1.1.5 Data visualisation
Data visualisation in ML is the process of graphically representing data to help understand patterns, trends and insights. It is invaluable at every stage of an ML project. It involves creating visual contexts such as charts, graphs, heatmaps and plots to make complex data more accessible and comprehensible.

During data exploration, visualisation helps to identify distributions, outliers and relationships within the data, guiding feature selection and engineering. In pre-processing, it aids in detecting anomalies, missing values and necessary transformations. During model evaluation, visual tools such as confusion matrices, receiver operating characteristic curves and precision-recall curves help to assess performance metrics and understand model behaviour.

Effective data visualisation ensures clearer communication of findings, enabling data scientists, stakeholders and decision-makers to grasp the implications of the ML models and the underlying data. Transforming raw data into visual formats simplifies the complexity inherent in large data sets and predictive models, facilitating better decision-making and more intuitive insights.

4.1.1.6 Typical methods of visualising data
Visual representation of data is one area where a focus on AI or ML can enhance how you organise and present data. The benefits of good visualisation must be a priority as, ultimately, we are judged by clients, stakeholders and/or users on the deployed AI or presentation of results. Visualisation is a fundamental enabler of learning from experience for everyone on the team. Figure 4.1 shows the stages of an ML project and highlights how visualisation can help with learning from experience at every stage (see Géron's eight stages in Section 4.2.2).

Figure 4.1 Visualisation is learning from experience at every stage of an ML project

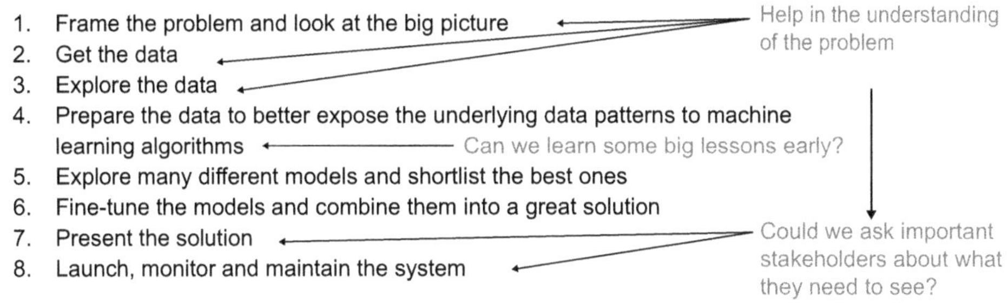

1. Frame the problem and look at the big picture
2. Get the data
3. Explore the data
4. Prepare the data to better expose the underlying data patterns to machine learning algorithms
5. Explore many different models and shortlist the best ones
6. Fine-tune the models and combine them into a great solution
7. Present the solution
8. Launch, monitor and maintain the system

Help in the understanding of the problem

Can we learn some big lessons early?

Could we ask important stakeholders about what they need to see?

If we can visualise data from the start of a project, we magnify our learning – a picture paints a thousand words. In addition, we are already organising our data so we can manage them and use them. If we are in a highly regulated profession, then early engagement with stakeholders and regulators might aid the acceptance of work and results. If we remind ourselves of the types of data we encounter – structured, unstructured and semi-structured – then we'll need to appreciate these types and how we visualise them.

It is worth noting that Python is the most common ML coding language of choice for computer and data scientists (Python, 2019). The reason for this is often given as the interactive nature of Python. In addition, we often see data pipelines in data visualisation software. Typically, Python is the scripting language within these software packages that automate standard tasks; SciKit-learn and TensorFlow have Python as a coding language (Scikit-learn.org, 2024). TensorFlow is also widening access to its ML libraries with other languages, such C++ and Swift (TensorFlow, 2019).

Some typical types of data visualisation are shown in Figure 4.2. The top row (a–c) are graphs of data that we might encounter in a spreadsheet. The middle row is an iso-contour plot, perhaps of the heights in a map. The image on the left-hand side is an iso-surface (d); from the iso-contours we can build the 3D iso-surface (e). With some vector calculus we can also calculate the gradients and perhaps find maximum and minimum heights. These ideas are very powerful if the data we are plotting are the learning rate of a NN or the profit from business activities. We can focus on the fast learning rate or maximise our profit. The bottom row of the examples (f and g) are networks; these are more complex structures to visualise. Here techniques such as stereo vision, augmented or virtual reality can help. Even large displays, such as power walls, allow a different perspective on data we are presenting.

These types of data visualisation techniques have been around for decades, and software on our computers enable us to create them. However, when we move to large data sets from simulations or big data, we need parallel data-processing capability. ParaView (www.paraview.org) is a parallel data visualisation ecosystem with Python at its core. It is built on the Visualization Toolkit open-source 3D graphics image processing and visualisation software library and constructed to run on parallel computers, so can process large data sets and produce graphics that an individual user can see. This is a major challenge when working with large data sets – working with large data sets on parallel computers is not something that should be attempted on your first project.

Most laptop and desktop computers are multi-core and multi-processor machines with graphics processing units (GPUs) that have enough processing power and capability to learn and develop ML/AI. Once tested on small hardware, you can then gradually use greater processing power on high-performance computing systems.

Figure 4.2 Standard data visualisation examples: (a) line graph, (b) scatter plot, (c) histogram, (d) contour plot, (e) 3D contour plot, (f) network diagram and (g) 3D network diagram

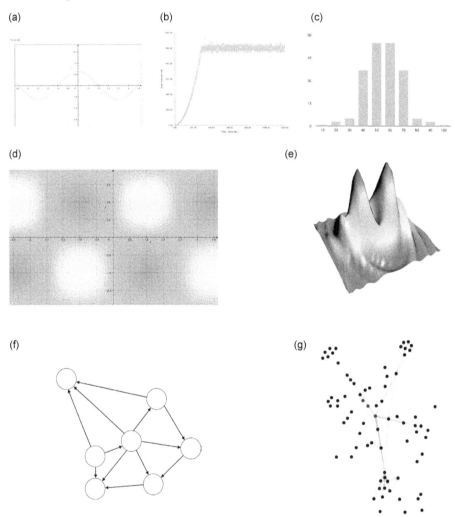

We are fortunate that most open-source software can be compiled, or is already compiled, for common operating systems. You can initially work on your own hardware using the same software you will use on larger hardware set-ups. We are also lucky to have both open-source and commercial software we can draw on. There are advantages and disadvantages to both, and you will need to consider what you will require. A search on your favourite search engine will bring up a list of the many data visualisation packages around.

We know that visualisation is an important aid to learning from experience, but there is an additional idea that we might be able to incorporate: can we create a learning environment? Can we develop a virtual reality (VR) or augmented reality (AR) simulated environment that we can learn from or humans can teach an AI system to learn from? We mentioned in the previous chapter that AI robotics gives us the opportunity to explore and utilise extreme environments. Could we build simulated environments to generate the data we need to design and operate our robots? Could we build a learning environment to let a robot learn how to do a task?

Blender is open-source software with a highly sophisticated environment for building simulations and presenting data (see Figure 4.3). The graphics are capable of producing photorealistic renderings and movies. This is often the base of computer games, movies and scientific visualisation. If we extend this to ML and AI, we could build a human and machine environment where the machine and humans interact to teach each other. Examples of this are found in engineering, where operators are taught how to operate machinery and pilots are taught how to fly in a safe environment. We are always guided by learning from experience, and the learning environment can be used to generate data.

Figure 4.3 Open-source software to build simulated environments – Blender (https://commons.wikimedia.org/wiki/File:Blender_2.90-startup.png, Creative Commons Attribution-Share Alike 4.0 International licence)

In medicine, AI can be used in diagnosis and treatment. Of particular importance in a surgical environment is what the surgeon sees. This can be developed in a digital environment first, then perhaps a physical environment and finally used on a patient. We can easily imagine a basic system, developing from scanning a patient to one where surgeons can work collaboratively with AI machines. Figure 4.4 shows examples of how AI is working in medicine from diagnosis to treatment. It shows robotic surgical equipment – the surgeon is potentially removed from the sterile operating theatre.

Figure 4.4 AI in medicine – humans and machines working together (Clockwise: https://en.wikipedia.org/wiki/Radiology#/media/File:Radiologist_interpreting_MRI.jpg, Creative Commons Attribution 4.0 International licence; https://en.wikipedia.org/wiki/Da_Vinci_Surgical_System#/media/File:Cmglee_Cambridge_Science_Festival_2015_da_Vinci.jpg, Creative Commons Attribution-Share Alike 3.0 Unported licence; https://commons.wikimedia.org/wiki/File:Cmglee_Cambridge_Science_Festival_2015_da_Vinci_console.jpg, Creative Commons Attribution-Share Alike 3.0 Unported licence)

4.1.2 Data quality

In this section, we start with the ever so familiar phrase about data:

Garbage in, garbage out.

With AI, and in particular ML, we need to perhaps change this to:

Good quality data in, garbage out.

We'll see later in the chapter that, although we might give our ML system good quality data to learn from, incorrectly used it can generate garbage. We should remind ourselves of the scientific method and be sceptical of our work and remember that it must be reproducible. This is particularly important for generative AI that is incredibly popular at the moment. This book is not a data management book and in a real-world project we must be very careful how we work with data. It is vital we are ethical and obey the law.

4.1.2.1 Data quality metrics

Five key data quality metrics are essential for ensuring the reliability and usability of data within an organisation: accuracy, completeness, consistency, timeliness and uniqueness.

- **Accuracy** measures how closely data values match the true values they represent, ensuring the data are correct and free from errors.

- **Completeness** assesses whether all required data are present, indicating that no critical information is missing.

- **Consistency** checks if data are uniform and standardised across different data sets and systems, reducing discrepancies and ensuring coherence.

- **Timeliness** evaluates the currency and relevance of data, ensuring they are up to date and available when needed.

- **Uniqueness** ensures that each data entry is distinct and not duplicated, maintaining the integrity and trustworthiness of the data set.

Together, these metrics provide a comprehensive framework for monitoring and maintaining high data quality, which is crucial for informed decision-making and operational efficiency.

4.1.2.2 Poor-quality data and its consequences

Poor-quality data can severely impact ML models, leading to a cascade of negative consequences. Inaccurate data introduce errors that can cause models to learn incorrect patterns, resulting in poor predictive performance and unreliable outputs. Incomplete data can skew the training process, leaving the model with insufficient information to generalise well to new, unseen data. Inconsistent data lead to difficulties in model training, as the algorithm struggles to reconcile conflicting information. Delayed or outdated data can make models irrelevant or inaccurate over time, especially in dynamic environments where timely insights are crucial. Additionally, duplicate data can inflate the significance of certain patterns, biasing the model. Overall, poor data quality undermines the integrity, accuracy and reliability of ML models, leading to misguided decisions, reduced efficiency and potential financial losses for organisations relying on these models for critical tasks.

4.2 DATA MANAGEMENT AND UTILISATION

Throughout the data lifecycle, there are various challenges and risks to consider, from how data are legally gathered and stored to ensuring they are processed in line with their intended use and are free from bias or misinformation. Figure 4.5 shows some of the typical legislation and regulation we must abide to when working with data.

Figure 4.5 Legislation and regulation

The UK General Data Protection Regulation (UK GDPR) and the Data Protection Act 2018 (DPA 2018) impose several legal restrictions on the use of data for AI applications.

These regulations aim to protect individuals' privacy and ensure the ethical use of personal data.

Here are the key legal restrictions and requirements:

1. Lawfulness, Fairness, and Transparency
2. Purpose Limitation
3. Data Minimization
4. Accuracy
5. Storage Limitation
6. Integrity and Confidentiality
7. Accountability
8. Rights of Data Subjects
9. Special Category Data
10. Data Protection Impact Assessments (DPIAs)
11. Data Protection Officer (DPO)

Practical steps for compliance include:

- **Conduct regular audits:** regularly review data-processing activities to ensure compliance with UK GDPR and DPA 2018.

- **Implement strong data governance:** establish clear policies and procedures for data management.

- **Conduct training and awareness:** train employees on data protection principles and the importance of compliance.

By adhering to these principles and regulations, organisations can ensure that their AI applications are compliant with the UK GDPR and DPA 2018, thereby protecting individuals' data rights and maintaining ethical standards. When using AI and ML, use the scientific method to systematically identify, analyse and mitigate data issues, leading to more robust, fair and accurate AI systems.

4.2.1 Risks in data handling

Using data in AI and ML poses numerous challenges and risks that can undermine the effectiveness of models and lead to unintended consequences. One primary challenge is ensuring data quality, as poor data can introduce biases, inaccuracies and inconsistencies that skew model outputs. Managing and processing large data sets also require significant computational resources and expertise, complicating the development and deployment of models. Additionally, the risk of over-fitting arises when models learn noise or irrelevant patterns in the training data, resulting in poor generalisation to new data.

Privacy and security concerns are paramount, as using sensitive or personally identifiable information necessitates stringent measures to prevent data breaches and misuse. Ethical considerations, such as avoiding discrimination and ensuring fairness, further complicate the use of data, especially when historical biases are embedded within data sets. Finally, regulatory compliance with laws such as GDPR adds another layer of complexity, requiring organisations to navigate legal frameworks while leveraging data for ML. These challenges and risks highlight the need for robust data management practices, ethical guidelines and continual monitoring to ensure the responsible and effective use of data in ML.

4.2.2 Preparing data for a machine learning project

Preparing data is an important aspect of all work with computers, not just AI or ML. In AI, learning from experience is paramount, and in ML we focus on learning from data. There is no guaranteed, surefire way to ensure your data preparation is right first time. You have to learn from experience and improve your data preparation with each iteration. So, when do we stop? That is something determined by the domain expert – this person sets the standard for what is fit for purpose. An Agile project style lends itself to this very well.

Data preparation is affected by numerous factors and is a science in its own right; it therefore needs experience to undertake. This experience could be learned during the project, or you could employ someone already experienced to do this for you.

> The Further Reading list at the end of this book has numerous ML examples that will give you a flavour of the data preparation needed.

We should keep in the back of our minds that if we are working with large data sets, we have the added complexity of obtaining the data, storing them, organising them, keeping the data safe and obeying any laws associated with them – a time-consuming task in itself. Aurélien Géron's eight stages of a machine project is a good place to start (Géron, 2017). These are:

'1. Frame the problem.
2. Get the data.
3. Explore the data.
4. Prepare the data to better expose the underlying data patterns to ML algorithms.
5. Explore many different models (or algorithms) and shortlist the best ones for the project.
6. Fine-tune your models and combine them into a great solution.
7. Present your solution.
8. Launch, monitor and maintain your system.'

Again, we notice immediately that there is considerable learning from experience involved in these eight stages. They are not sequential, and we may revisit or loop over stages 1–5, 1–7 or 1–8 – going through stages 1 to 8 in one pass will be the exception rather than the rule. We might use smaller data groups so we can learn quicker because we are not processing lots of data. This can be helpful when assessing multiple algorithms, as there is more work associated with this. Some learning can be gained by working on smaller subsets at an early stage. It simply reduces the time between iterations.

If we have the luxury of a large data set, we will need to break up the data into training data and test data. This is important for the trade-off, where we want to trade off complexity and error; this is covered in more detail in Section 4.2.3.9. We might then break up our training data into data bins of different sizes and use those different sized bins to test our algorithms. As we refine our approach, we can then use the test data to see if our learning works on data that have not been seen by the AI or ML system before.

Test data must not be used for training your AI or ML system. One of the reasons for doing this is that the algorithm will use the test data while training, so we already know how well the algorithm deals with these data. It is essentially a waste of time because we are not adding further information to the system. Independent test data give us an independent test of the ML we have done.

Once we have our data and they are divided up into appropriate bins, we can then start to clean or scrub the data. This means that we prepare our data in such a way that they give our algorithms the best chance of success.

In AI, we can encounter many data types, such as numbers, strings and Booleans, or they could be more sophisticated data, such as audio files, movie files, script, text, documents and, perhaps, signals from actuators, sensors – we might even have to generate our own data in simulations. Digital twin is an exciting development that has gained momentum over the last 20 years or so. Data will come to us in many forms, in varying quantities.

In AI, we will need to work with analogue data. Here we'll typically employ analogue to digital converters so we can represent an analogue signal as a series of discrete numbers that can be understood by a digital computer.

Part of the preparation of data is data cleaning or data scrubbing. As we are going through the stages of a ML project, we must always be aware of the quality of our data preparation. Poor data preparation will give spurious ML results or misrepresent the data we are trying to analyse.

Scrubbing can include:

- reducing the volume of the data we are working on, perhaps removing irrelevant data;
- reformatting data;
- removing incomplete or duplicated data;
- dealing with missing data;
- generating data;
- scaling data so algorithms perform well.

There are many tools and techniques for data preparation and standard libraries to help us, and we need experience to do this well.

As important as data preparation is data visualisation. Visualisation of data (see Section 4.1.1.5) will also affect how you prepare your data. This is often overlooked until we try to visualise a data set that spans multiple processors on a large supercomputer.

4.2.3 Algorithms that learn from data

Algorithms are at the core of problem-solving, engineering and computer science. In AI, we are interested in how they can help us with learning from experience; how they can help us to create intelligent entities, products and services that can learn and improve. Algorithms are studied academically, and this can lead to a lifetime's pursuit into complex and rich research. Algorithms are also simple procedures we use to teach ourselves how to program a digital computer (from simple flowcharts to complex event-driven object-orientated software). Narrow AI, and in particular narrow ML, give us a chance to understand how an algorithm can learn from data. The NN and its many variants teach themselves how to solve a problem. It is an algorithm that learns its own algorithm to, say, play chess.

We should note that an AI or ML algorithm may not always be the best option to solve our problems. Scientists, mathematicians and engineers will often have more elegant and efficient ways to solve problems. However, we don't always have this capability, and here ML and AI can help us, especially where we have a lot of data to work with (e.g. data from telescopes, simulations, sensors and so on). Also, algorithms can be expressed in a multitude of ways, as computer code, flowcharts, mathematical notation, pseudocode and so on. For those that want to delve deeper into algorithms, Tom Mitchell's book, *Machine Learning* (2018), provides an excellent introduction to many algorithms and their practical uses, including those that we describe below and their derivations.

AI and ML are, at their fundamental core, based on the scientific method, and we must use these techniques with care. Our learning problems are non-linear, iterative and probably statistical in nature. We must use algorithms cautiously because they are often general approaches to problems – domain specific techniques may well be far better suited to our problems. Part of our approach to using ML and AI is to build our learning and experience of what algorithms can achieve.

ML and AI draw on techniques from well-established academic disciplines such as operational research, control theory, mathematics, numerical analysis and engineering. For example, decision trees are frequently associated with ML, but they have been at the heart of safety engineering and management theory for decades.

In this chapter we aim to introduce what an algorithm is and why algorithms are so important to ML and AI. Algorithms have their advantages and disadvantages in terms of accuracy, performance and processing time. This chapter also includes an explanation and examples of a few of the most common algorithms in ML. We will explain their use and the challenges that are encountered in deciding which to use and why.

4.2.3.1 What is an algorithm?

The definition of an algorithm is a process or set of rules to be followed in calculations or other problem-solving operations, especially by a computer. Simply put, an algorithm is a set of instructions designed to perform a specific task. This can be as simple as the process we follow to multiply two numbers together, something that we are taught at an early age and take for granted. Humans do this in decimal, and computers in binary. It is only when you have to explain what you are actually doing that it seems complicated – we need to be explicit because our digital computers need explicit instructions. An algorithm could also be something far more complex, such as compressing a computer file or playing a compressed video file.

We have previously covered learning types and data types, but it is worth addressing the basic concepts again before we go any further into the topic of algorithms.

In a supervised learning model, an algorithm learns on a labelled data set, which provides an answer key that the algorithm can use to evaluate its accuracy on training data. In an unsupervised model, it uses provided unlabelled data that the algorithm tries to make sense of by extracting features and patterns on its own. In both semi-supervised and reinforcement learning models, data can be both labelled and unlabelled.

Basic algorithm types include:

- **Regression** – generally used for forecasting and finding out the cause-and-effect relationship between variables.
- **Classification** – where input data can be separated into groups, for example male or female.
- **Clustering** – the grouping together of data that are similar, for example grouping of images of cats, dogs and rabbits based on data attributes.
- **Association** – here features of data can be associated with other features and, in an unsupervised learning condition, can then associate other data, for example people who generally buy bread and butter also typically buy milk.
- **Control** – where the error measurement between a set point and a measured value will help to decide what to do, for example an autonomous vehicle controller can either apply controlled braking or a collision avoidance manoeuvre.

In Table 4.1 we show a broad correlation between learning models, algorithm types and example algorithms, some of which will be described later in this chapter.

4.2.3.2 What is special about AI/ML algorithms?

Generally, an algorithm takes some input and uses mathematics and logic to produce the output. In contrast, an AI algorithm takes a combination of both – inputs and outputs – simultaneously in order to 'learn' from the data. In ML, once it has been trained, the ML algorithm can produce outputs when given new inputs.

Table 4.1 Examples of the types of learning and algorithms

ML categories	Algorithm types	Example
Supervised learning/labelled data	Regression	Decision tree
		Linear regression
	Classification	SVM
		KNN
		Naive Bayes
Unsupervised learning/ unlabelled data	Clustering	K-means
	Association	Apriori
Semi-supervised learning/mix of labelled and unlabelled data	Classification	Semi-supervised SVM
	Clustering	K-means
Reinforcement learning	Classification	Ensemble learning – decision forest
	Control	Q-learning

4.2.3.3 What are self-learning algorithms?

A self-learning algorithm is programmed to refine through iteration its own performance; that is, it learns from itself to improve its accuracy in its ability to perform. In the context of ML, this often requires considerable computational resource. It can be best described as a system into which you feed your requirements (i.e. the desired outcome plus various parameters) and over time the outcome is achieved. An example is recommendation engines where the system gets better and better at recommending 'things', which people then purchase as more data points are processed.

4.2.3.4 What are the algorithms used in ML and AI?

ML and AI algorithms include:

- **Linear regression** – this is a model that assumes a linear relationship between the input variables (x) and the single output variable (y). More specifically, that y can be calculated from a linear combination of the input variables (x). For example, imagine arranging a number of blocks by weight variable y when you cannot weigh them and only know their x (length, height and width) variables. A single input is known as simple linear regression and multiple input is known as multi-variable, multi-variate or multiple linear regression. One use could be to predict a human's ideal weight based on height, build, gender and so on.

- **Logistic regression** – this is a mathematical model used in statistics and ML to estimate the probability of an event occurring having been given some previous data. Logistic regression works with data where either the event variable (y) happens (1), or the event does not happen (0). An example could be around credit card transaction approval when multiple inputs such as time of purchase, place of purchase and type of purchase may determine if the transaction is approved or not.

- **Decision tree** – this is one of the most common ML algorithms in use today. It is a supervised learning algorithm used for classifying problems by moving down a tree root from node to node testing an attribute at each node. It is typically used in expert or smart systems: it can advise a course of action to be taken based on previous successful actions, and is typically seen as a series of questions and suggestions culminating in a 'was this information useful?' question on self-help computer systems. An example could be the fault light on a non-functioning printer: is it red or green? If green, then check if there is paper in the input tray, if the answer is yes, then check for a paper jam and so on, until the problem is identified. The learning element of the algorithm can weight suggestions based on previous successes with particular attributes. Equally, decision trees can be used in association to classify patients and likely disease based on health questions, or loan applications and previous credit history.

- **Random forest** – this is an ensemble learning method for classification, regression and other tasks. It operates by constructing a multitude of decision trees at training time and outputting the class that is the mode of the classes or mean/average prediction of the individual trees. With that said, random forests are a strong modelling technique and much more robust than a single decision tree. Random forests aggregate many decision trees, which limits the over-fitting problem of deep decision trees as well as error due to bias, and therefore the ability of the system to give useful and meaningful results. The use of random forest adds additional randomness to the model, while growing the trees. Instead of searching for the most important feature while splitting a node, it searches for the best feature among a random subset of features.

A commonly used example to explain the difference between decision trees and random forests is that in a decision tree a person may ask one friend to recommend a holiday destination based on a number of questions about preferences and previous holidays, whereas in random forest mode a person may ask the advice separately of multiple friends who again ask a series of different random questions based on their individual likes/dislikes and previous experiences of different holiday experiences.

- **KNN** – this is a supervised learning algorithm. This means that we train it under supervision and using the labelled data already available to us. It is a relatively simple algorithm used by many organisations and software tools – simple in that it assumes that similar things exist in close proximity to one another. It is typically used in recommendation applications or 'more like this' type systems such as Amazon recommending books or Netflix recommending films based on a particular genre, category, rating, lead actor, director and so on.

The classic travelling salesperson problem is often solved using a nearest neighbour heuristic, and is one of the first algorithms that comes to mind in attempting to solve this particular type of location/route problem. This classic problem is one in which a salesperson has to plan a tour of cities that is of minimal length. In this heuristic, the salesperson starts at some random city and then visits the city nearest to the starting city and so on, only taking care not to visit a city twice. At the end, all cities must be visited and the salesperson must return to the starting city. Use of the KNN algorithm gives a quick solution compared to a brute force approach where every permutation is calculated. It should be noted, however, that KNN may not always give the optimal route, especially if there are many data points. The KNN can be applied to many other movement problems, such as moving a robot or planning the router of a machine tool.

- **SVM** – this is a supervised ML algorithm that can be used for both classification and regression challenges. An SVM model can be best described as points in space mapped into categories separated by gaps that are as wide as possible. New data points then fall on one side of the gap and are then placed into that category. SVMs have many uses, ranging from image recognition to handwriting recognition to satellite data classification. A simple example may be categorising images of cats and dogs where an image must fall on one side or other of the gap.

  When supervised learning is not possible due to unlabelled data, then an unsupervised approach needs to be taken. SVM attempts to find natural clustering of data into groups and new data will then fall into one of these groups.

- **Naive Bayes** – this is a classification technique based on Bayes' theorem with an assumption of strong independence among predictors. Simply put, a Naive Bayes classifier assumes that the presence of a particular feature in a class is unrelated to the presence of any other feature. It is used in predicting membership probabilities for each class, such as the probability that a given record or data point belongs to a particular class. A simple example could be the prediction of the probability of you having a particular illness or disease based on the data recorded about you, which may include ethnicity, age and gender, and not just a list of dependent symptoms such as a rash or high temperature.

- **K-means** – this is a centroid based clustering algorithm, which means that data are clustered around a centre. K-means is an iterative algorithm, and has two unique steps: the first being a cluster assignment and the second being a move to the centroid step.

  Initially, you must select a number of clustered centres depending on the number of clustered groups you want to create. Measurements are then made from the various data points through each of the data points and, depending on which cluster is closer (the cluster A centroid or cluster B centroid or cluster C centroid and so on), the algorithm assigns the data points to one of the cluster centroids. K-means then moves the centroids to the average of the points in a cluster. In other words, the algorithm calculates the average of all the points in a cluster and moves the centroid to that average location. The two steps are then repeated until an end condition is met.

  A real-world example of this algorithm is in the segmentation of customers for marketing purposes into various persona categories to allow better targeted marketing messages to be sent to the different personas, thereby helping to increase engagement or sales. In marketing you would typically try to limit the number of personas to a manageable number – let's say six would be optimum – but how we determine that for our particular industry or market segment requires some learning from experience.

  The choice of the number of clusters to determine the optimum number of clusters (i.e. the 'k') can be quite complex and may be decided by adding an additional cluster until it no longer makes a significant difference. A single data point could also be considered a cluster. The decision criteria for k is beyond the scope of this book. Again, we see the importance of learning from experience while we are building our AI or ML technique.

The examples above are the tip of the iceberg in terms of the algorithms used in ML and AI, where there are literally thousands, if not millions, used for specific industries, for solving specific problems or within individual applications.

4.2.3.5 What is a deep learning algorithm?
As with the algorithms described in the last section, there are also a number of algorithms used in deep learning on NNs. The detailed workings of the algorithms are beyond the scope of this book other than to provide a description of their function and use. These algorithms include:

- **Multilayer perceptron neural network (MLP NN):** this is a class of feedforward artificial neural network (ANN). MLP utilises a supervised learning technique called backpropagation for training. Its multiple layers and non-linear activation distinguish MLP from a linear perceptron. It can distinguish data that are not linearly separable.

- **Backpropagation:** this is short for 'backward propagation of errors', and is an algorithm for supervised learning of ANNs using gradient descent. Given an ANN and an error function, the method calculates the gradient of the error function with respect to the NN's weights. Weights can then be adjusted to improve the error.

- **Convolutional neural network:** CNN is a deep learning algorithm that can take in an input image and assign importance, based on learnable weights and biases, to various aspects/objects in the image and be able to differentiate one from the other. The CNN was inspired by the human brain's processing of visual data. These NN are more versatile and can be used on data other than images (e.g. audio, self-driving cars and more).

- **Recurrent neural network:** RNN is a class of ANN that processes a sequence of data along a timeline and remembers 'in memory' the sequence of data or a variable amount of data to use later in its processing. There is a connection between the data points, just as we have when we read individual words and link them together in a sentence. RNNs have been derived from feedforward NNs. They are typically used in speech recognition and connected handwriting recognition applications where what has come previously may influence what comes next or how it is interpreted.

- **Generative adversarial network:** GAN is a class of ML frameworks designed by Ian Goodfellow and his colleagues in 2014: two NNs contest with each other in the form of a zero-sum game, where one agent's gain is another agent's loss. Apart from gaming, GANs are also used in image improvement and event generation and are used in the generation of 'deepfake' photorealistic human images resembling real photographs and videos generated from a single image.

- **Long short-term memory (LSTM):** this is an artificial RNN architecture used in the field of deep learning. Unlike standard feedforward NNs, LSTM has feedback connections. It can not only process single data points (such as images), but also entire sequences of data (such as speech or video).

- **Restricted Boltzmann machine (RBM):** this is a restricted two-layer ANN with a visible and a hidden layer. Originally invented in the 1980s, and still used today, it can learn probability distribution from its data inputs. RBMs can also be trained in either supervised or unsupervised ways, depending on the task. RBMs are used in topic modelling, classification, collaborative filtering and feature learning, and have found uses in other applications such as speech recognition. The RBM is the basis for DBN.

- **Deep belief network (DBN):** this can be created from multiple layers of simple, unsupervised networks such as RBMs. The DBN in its simplest form is composed of many layers of latent variables, with connections between the layers but not between units within each layer. When the DBN is initially trained without supervision on a data set it can learn to construct its inputs probabilistically. After initial training, supervised training will then allow it to perform classification. DBNs are often used in the field of drug discovery.

Many other deep learning algorithms are available, but the descriptions above show that significant skills, knowledge and experience are required to choose, apply and make successful use of these algorithms. Each type of algorithm described above is a rich area of mathematics and theory. They are all rooted in the scientific method, and a rigorous approach to the use of these algorithms is likely to be found. What we mean here is that, although the interpretation of a complex algorithm is difficult, its roots in the scientific method will help to build a professional and rigorous justification for its use and results. Deep learning approaches contain more data and steps than a human is capable of comprehending. Again, we should remind ourselves that these general learning techniques may not be the optimum solution to our problems. Here, the domain expert will play a pivotal role in ascertaining what is fit for purpose. They will also be able to help with heuristics that are so vital to ML.

4.2.3.6 Neural networks – a closer look
Here we take a closer look at the NNs. In particular we describe the following:

- the link between the NN and the AI agent;
- the basic building blocks of the NN;
- building algorithms using reinforcement learning; and
- a few basic examples of what a NN can do.

The NN gives machines the ability to learn from experience and develop their own algorithms (it is not easy to interpret what these algorithms are, as they are encoded in the NN). In recent years, our understanding on how a NN works has made significant progress. Progress includes winning complex strategy games such as chess and Go, as well as the perhaps more challenging tasks of controlling complex machinery such as robots. In doing so we are giving machines more autonomy.

There are many types of NN available to us, and one that stands out as having made a significant contribution to ML is the CNN. However, we are not going to go into detail about this type of NN here because the details are more complicated than the discussion we intend to have at this point.

To learn from problems that are complex and perhaps random (statistical), we need to use deep neural networks (DNNs). Typically, we employ DNNs for problems such as robot control, image analysis (e.g. medical images) and adversarial game playing. We will introduce what these are and why this makes training them iterative and requires specialist hardware (e.g. high-performance computers, GPUs, etc.) as we progress through the chapter. A DNN has many layers making up the NN; this is why it is called a deep NN.

4.2.3.7 The link between NNs and the AI agent

The NN is often shown as a sub-topic of ML. The development of the NN started in the 1940s and is based on the biological nature of the human brain. In 1943 Warren McCulloch and Walter Pitts introduced the concept of a neuro-logical network by combining finite state machines, linear threshold decision elements and memory (McCulloch and Pitts, 1943), followed in 1947 by a paper extending these ideas to the recognition of patterns (Pitts and McCulloch, 1947). The first reinforcement NN built by Marvin Minsky occurred in 1951. This was an analogue or electronic device that used reinforcement to learn (Minsky and Papert, 1988). This was the start of adaptive control, and we see a close link to engineering emerging. In 1962 Frank Rosenblatt named and defined machines called 'Perceptrons' (Rosenblatt, 1962); however, it took until the 1980s for NNs to gain significant traction. Today the theory is understood in greater depth. Digital computers have facilitated the use of CNN to undertake sophisticated tasks such as learning to control robots and win combinatorically complicated games.

The learning AI agent needs the ability to learn. Reinforcement DNNs or CNNs do just that. They can learn from sensors and actuators or from well-engineered computer simulations or games.

Narrow AI has been focused on specific problems. The NN is a more generic learning technique that can be applied to a wider range of scenarios. Training NNs is not easy, and understanding their output is difficult also. This poses challenges for justifying them and using them.

4.2.3.8 The basic building blocks of the neural network

In this section we introduce the concept of learning that is fundamental to NNs. The perceptron was the initial concept from which our more sophisticated NNs of today evolved. As we build up the basic building blocks of the NN, we will see how the perceptron fits in.

The NN has a basic mathematical schematic, shown in Figure 4.6 and based on the original work of McCulloch and Pitts in 1943. In this figure, we also see a simple representation of a human brain neuron.

Each of these NN neurons form layers (lots of neurons), and multiple (hidden) layers make up a network. The first layer is the input layer where data are fed into the network. The data can be output from a simulation, the colour of pixels in an image or data from sensors. The data pass through the network, and as they do, just like the human brain, neurons are activated (or fire). A simplified three-layer version of a NN is shown in Figure 4.7.

Figure 4.6 The schematic of a neural network neuron and a human brain neuron
(McCulloch and Pitts, 1943; https://commons.wikimedia.org/wiki/File:Neuron3.png, Creative Commons
Attribution-Share Alike 3.0 Unported licence)

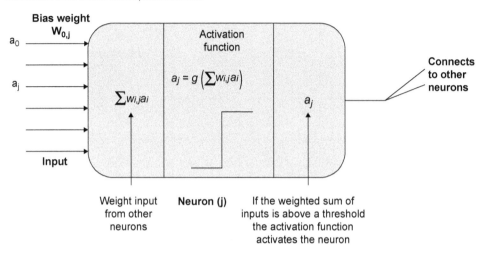

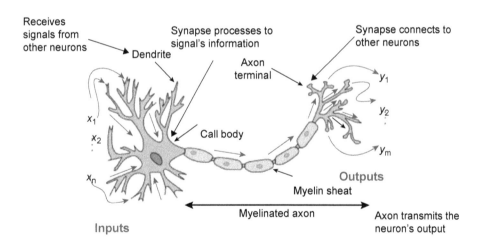

There can be any number of layers, and DNNs have lots of layers. Unfortunately, we do not know what number of layers, or depth of DNN, we need before we start training the NN. That is something we learn from experience as the trainer, and part of the hyper-parameters that we will systematically learn from. In this description we have been careful to use only feedforward networks where the input of the node is passed onto the next layer. RNNs also allow nodes to communicate with other nodes and themselves on the same layer. These types of network are described as more dynamic and can behave chaotically. We will stick with the feedforward NNs here, as we are building an intuition or understanding.

Figure 4.7 The schematic of a neural network with an input, hidden and output layer. In NNs, a neuron is a node in the network and the node is made up of an input and activation function and an output. Nodes are connected via links (https:// commons.wikimedia.org/wiki/File:Colored_neural_network.svg, Creative Commons Attribution-Share Alike 3.0 Unported licence)

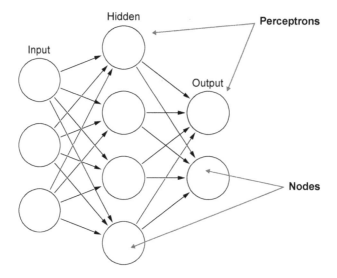

NNs work by taking the output (if it is activated) from nodes and passing this information via a link. The link connects the output of a node to nodes on the next layer. Each link from the output of a node is uniquely weighted as it is passed to other nodes. We do not know what these weights are, and training is about determining what the weights should be.

An example might help here. Say we have a high resolution picture of a scene taken on your smartphone – the type of picture doesn't matter. We can analyse this picture using a NN. The picture is held on your phone as a matrix of numbers that represent the colour of the image at a particular location. The NN takes this matrix as its input and passes this number to every node on the first layer. Before it does so, it weights it by a unique weight between the input node and the layer node it is going to. Each node on the first layer then adds up the colour number and weights it has received. This sum of weighted input colour numbers is then passed through the activation function. If the output of this operation is greater than a set value, the node is assumed to be activated and the result is passed onto the next layer. If it is not activated, then this node is not activated, and its output is zero. The layer of now activated or inactivated nodes is fed into the next layer and the process is repeated. The activation function is trying to represent how our brain fires its neurons when parts of our NN are stimulated. As information is passed through the network of nodes, nodes are fired or not fired. This is a very simple analogy, and training a network to find the best weights needs a bias node as well. This helps us to shift the threshold of nodes activating or not.

Whether a node is activated or not is determined by the activation function. The activation function uses the input function as its input. This input function is the sum of the output of nodes connected by the links multiplied by the weights of the links between the nodes. We then need to determine if the node is activated. If a hard threshold (activation function output is above a set value) is used, then this is a perceptron. These types of NN can represent non-linear problems that are Boolean in nature. The sigmoid perceptron uses a differentiable sigmoid activation function, and this means DNNs can now represent arbitrary functions, again non-linear. This makes them very powerful. The threshold activation function is a hard step change between the neuron activating or not. The sigmoid (logistic) activation function is differentiable and is called softer because the decision to fire the node is not a step change. There are other types of activation function, for which use is determined by the experience of the trainer.

The trainer needs to train the NNs to determine the unknown weights of each of the links. This is not a simple task, and a role of the future for AI personnel. Training a NN involves adjusting weights in order for the NN to improve. Techniques involve forward and back propagation to update weights based on the error between the desired output and the NN's output. Knowledge of vector calculus will help here, and a NN domain expert will help too. Up to now we could simplify the NNs as ones that can understand functions, given examples of what those functions are. We could use a NN to learn from examples that we know. This makes the error known. What happens when we do not have labelled examples to learn from? We could build learning examples and try to form labels from our understanding of the NN.

What happens if our problem has lots of unknown combinations – such as a game with many possible moves? We simply may not have the data to understand all the possible moves. We need to measure, somehow, the performance of the NN. This is called reinforcement learning, and the NN can learn by itself. Reinforcement learning fits naturally with the learning agent schematic. The agent has a learning component, and the learning component tries to maximise its reward or minimise its loss. Norvig and Russell (2016) note that reinforcement learning might be considered as encompassing all of AI. In an environment, an agent will have a percept and will learn via the learning element to achieve a goal. It will use its actuators to act on the environment. This elevates products to intelligent entities and gives products such as robots autonomy. An agent using a NN and reinforcement learning has to learn how to achieve a goal in an environment. This learning is complex. Somehow, we need to train the NN to explore its environment, create a policy to achieve a goal and understand what it can do to the environment with its actuators.

Examples of reinforcement NN successes are:

- AlphaGo, the computer that was taught to beat the world Go champion (see Section 1.1.2.1). The computer was so successful, it used moves that impressed Go masters. A supercomputer played itself at the game of Go to learn how to play. This is an engineered simulation environment.

- OpenAI, who used a NN to teach a robot how to solve a Rubik's Cube (OpenAI, 2019). The robotic hand can manipulate the cube to solve it because it has many sensors and actuators. The NN learned how to solve the Rubik's Cube and also how to manipulate its environment.

4.2.3.9 Hyper-parameters, algorithms and the trade-off – under- and over-fitting

Our learning from experience involves assessing the efficacy of an algorithm to test our hypothesis. We will try different types of algorithm and adjust the algorithms' tuning parameters to get them to perform the best they can. The tuning parameters are called hyper-parameters, and they are used to set up our algorithms; or, to put it another way, they tell the algorithm how to learn. Examples of hyper-parameters include layers in a decision tree or NN, the order of a curve in regression, the number of nodes in a NN, the type of activation function or the accuracy of a classification.

We will not know before we start what our hyper-parameters need to be to obtain the most effective algorithm. This is part of our learning from experience as we build our ML system. There are also many ways to assess the efficacy of an algorithm. The most common one used is the trade-off between complexity and accuracy. To do this, we need to understand under- and over-fitting.

These are exactly as their names imply: over-fitting the model over-fits the data too well, and under-fitting under-fits the data and does not capture the true nature of them. You can reproduce over-fitting and under-fitting on a spreadsheet by simply fitting different curves to a data set.

With over-fitting we have a data set that represents a linear straight line (see Figure 4.8). This could be from experimental measures of temperature through a metal bar. The good fit straight dashed line in the graph shows the best fit to the data set. There is only a small variation of the data when compared to the best fit straight line. If we try to be too clever and use a more complex mathematical fit, say a parabola or a higher order polynomial, then we will begin to see larger errors. In the curve fitting example, the higher the order of the curve, the more oscillatory the data fit we obtain. These show up as large errors away from the actual data points we use to train our ML solution. In fact, if we have nine data points, we need a ninth order polynomial to obtain an exact match of the data points with our polynomial. However, this ninth order polynomial could oscillate widely and be a very poor fit away from the data points. To recap, over-fitting means our model or algorithm is too complicated for the data we are working with.

Under-fitting, on the other hand, is the opposite; it tries to fit a curve (using our simple curve idea) that is not complicated enough. In Figure 4.9 we see that our data set is well represented by a parabola (dashed line) or a second order curve. If we try to fit a simple linear straight line to the data, we again see large errors.

So, our algorithm suffers from two types of error: bias error and variance error. The bias is how far your prediction is from the actual data. Variance is how scattered around our prediction is. With over-fitting, we see a high variance around and low bias with our training data. With under-fitting, we see a low variance around and high bias with our training data. The trade-off, shown graphically in Figure 4.10, minimises the error by finding the best level of complexity. This is not the only way to assess the efficacy of an algorithm. There are many other ways, and there are numerous examples to be found in the books listed in Further Reading at the end of this book. We should also note in passing that the trade-off uses test data that have not been used to train our algorithm.

Figure 4.8 A simple example of over-fitting

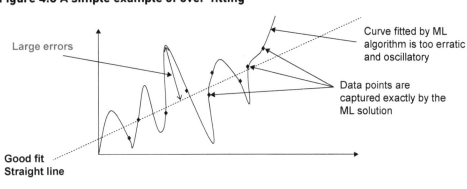

Figure 4.9 A simple example of under-fitting

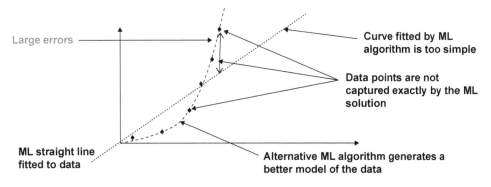

Remember that the learning from experience within a ML project involves the whole team. The domain expert is the person that defines what is fit for purpose.

4.3 GENERATIVE AI

Generative AI represents a groundbreaking advancement in the field of AI, characterised by its ability to create new content, whether it's text, images, music or even entire virtual environments, that closely mimics human creativity. Unlike traditional AI, which is often designed to recognise patterns and make predictions based on existing data, generative AI leverages complex algorithms and NNs to generate novel outputs. This technology has transformative potential across various industries, from entertainment and art to healthcare and scientific research, by enhancing innovation and automating complex creative processes. As it continues to evolve, generative AI is poised to redefine the boundaries of what machines can achieve, offering both exciting opportunities and ethical considerations for society.

Figure 4.10 Algorithm assessment, the trade-off and finding the balance between under- and over-fitting

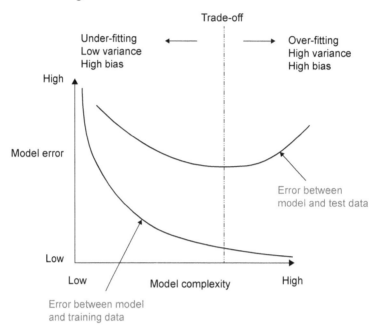

4.3.1 Key generative AI (Gen AI) terms

If we are interested in Gen AI, these are typical terms we should be familiar with:

- **GANs:** an ML framework with two NNs – the generator, which creates data, and the discriminator, which evaluates it – working together to produce realistic data.

- **VAEs:** a type of NN that learns to encode data into a compact form and then decode it to generate new, similar data.

- **Transformer models:** advanced NN architectures, such as a generative pre-trained transformer (GPT), that are particularly good at generating human-like text.

- **Deepfake:** synthetic media, especially videos or images, created using AI to mimic real people, often raising ethical concerns.

- **Neural style transfer:** a technique where AI applies the artistic style of one image (e.g. a painting) to another image.

- **Latent space:** A lower-dimensional representation of data that captures its essential features, allowing for the creation of new data points by sampling from this space.

- **Prompt engineering:** crafting specific input prompts for generative AI models to guide them in producing desired outputs.

- **Tokenisation:** the process of breaking down text into smaller units (tokens), such as words or subwords, which are then used by AI models for processing and generation.
- **Training data:** the data set used to train generative AI models, which they learn from to produce new content.
- **Inference:** the phase where a trained AI model generates new data or makes predictions based on new input data.

4.3.2 Purpose and use of generative AI

If we have a human-centric ethical purpose, harnessing generative AI involves a few key steps:

1. **Train the model:** we start by teaching the AI using lots of examples. For instance, if we want it to write stories, we give it many stories to learn from.
2. **Create new content:** once trained, we can ask the AI to create new content. For example, we might give it a few words and ask it to write a paragraph, or show it a picture and ask it to create something similar.
3. **Fine-tune with prompts:** we can guide the AI to produce specific results by giving it detailed instructions or examples of what we want.
4. **Use for specific tasks:** generative AI can be used for various purposes, such as making artwork, writing articles, designing products or even creating music.
5. **Improve over time:** we can keep improving the AI by giving it more data to learn from and adjusting its training to make it better at generating what we want.

In simpler terms, we teach the AI by showing it lots of examples, ask it to create new things based on what it learned, guide it with instructions and continuously improve it to make it more useful for different tasks.

4.4 SUMMARY

This chapter delved into the relationship between data and AI, focusing on how various data types contribute to AI and ML projects. It introduced key concepts, such as structured, semi-structured and unstructured data, and emphasised the importance of handling big data, its volume, variety, velocity and veracity. Additionally, the chapter explored data visualisation techniques and their critical role in every stage of a ML project, aiding in understanding patterns and improving decision-making.

The chapter discussed the significance of data quality, highlighting how poor-quality data can lead to unreliable AI models. Key data quality metrics, such as accuracy, consistency and timeliness, were explained to ensure that AI systems perform effectively. It also underscored the legal and ethical responsibilities tied to data management, particularly in light of regulations such as GDPR, and outlined the risks involved in handling large data sets.

The chapter introduced a step-by-step approach in preparing data for ML projects, from framing the problem to data cleaning and visualising the results. It stressed the iterative nature of ML projects, emphasising that learning from experience is central to success.

The chapter then explored various ML algorithms, including supervised, unsupervised and reinforcement learning techniques. It provided examples of algorithms, such as regression, decision trees and support vector machines, explaining their uses and challenges. Finally, it covered deep learning algorithms, such as neural networks and reinforcement learning, which empower AI to learn autonomously from complex data, leading to advanced applications like AlphaGo and robotics.

Generative AI, a groundbreaking AI field, was discussed, showing how it creates new content such as text, images and music. The chapter explained key terms and techniques such as GANs and transformer models, highlighting the immense potential and ethical considerations of generative AI in creative industries.

5   AI IN ORGANISATIONS

This chapter looks at the simple ways in which AI is an opportunity for an individual, a team, an organisation and in society as a whole. AI, and in particular ML, is helping us daily. It's worth identifying how you are using it today (smartphones, smart TVs, smart roads, search engines). Paul R. Daugherty and H. James Wilson have written about their experience doing just that as management consultants in *Human + Machine: Reimagining Work in the Age of AI* (2018). In their book they observe how AI is applied in many projects. They describe human-only systems, machine-only systems and human-plus-machine systems. They paint a picture of how the future may look, what roles humans and machines will play and what humans and machines will do together.

5.1 AI OPPORTUNITIES AND BUSINESS CASES

Ever since the realisation that artificial intelligence was possible, we have considered the pros and cons of exploiting AI. It is only in the last couple of decades, when it has become technologically possible on a large scale, that we have seriously considered the challenges, issues and ethics around it; in particular, intelligent learning machines replacing humans in various roles. In fact, in some industries, automation has been a challenge since the first industrial revolution. In other industries and professions, it will be a long time until a machine will replace a human and, in certain roles, it will likely never be the case. It is likely that in many aspects of our personal and working lives we will not be replaced by AI, but work alongside learning machines in some way.

5.1.1 Opportunities for AI in organisations

In the simplest sense, AI can augment humans and machines and also augment how humans and AI machines can work together. This ties in nicely with Stuart Russell's solution to the control problem of AI; we allow the AI to help us to achieve what is best for humans (Russell, 2019).

5.1.1.1 Augmenting humans and machines
If we take the next step from ML to thinking about humans and machines together, we now need to consider how humans can augment machines and how machines can augment humans. We are at a point where the ethical purpose of our AI system is important. Roles for humans and machines will certainly emerge to ensure our AI systems have a human-centric ethical purpose.

5.1.1.2 Augmenting humans

Firstly, let's look at ways in which machines can augment human capability through the effective use of information technology. Improvements can range from slight increases in performance right up to making us superhuman.

To do this we need to introduce a new concept: intelligent automation or augmentation (IA). IA aims to use ML technologies to assist rather than replace humans. IA can take technology such as computer vision, NLP and ML and apply it to RPA, allowing the automation of processes that don't necessarily have a rules-based structure, for example using an IA system alongside an oncologist to identify cancer cells in biopsy slides – previously the task would have required two oncologists. This is an example of a human and machine working alongside each other for the benefit of humankind.

Recent advances in IA technologies have helped to advance human potential by increasing worker productivity, alleviating mundane and repetitive tasks and introducing and enhancing convenience in our lives. For decades, some machines have been much better than humans at performing certain tasks, but what we increasingly see now is heightened penetration and a continuous acceleration of IA across many industries and spheres of life. Over the next decade, we will see further rapid advances along these fronts as technology improves and acceptance of IA increases. We can already begin to see, learning from experience, that AI's technological disruption of knowledge workers is redefining their jobs. Justified trust is key to adoption and exploitation of this technology; it requires good governance as well as aligned assurance regimes.

5.1.1.3 Augmenting machines

The key areas where humans augment machines are in making AI and IA more understandable and acceptable to people in general.

At a simple level, machines can increase our learning and make it more effective. We humans have to orchestrate that learning and provide support when the machine hits a contradiction or fails. We also need to ensure that machines are doing the right things. Remember, we need human-centric ethical purpose to our AI systems. Machines don't have consciousness or emotion, which makes empathy a big problem. Humans will need to set the goals not just about where we are going but also how we are going to get there. An example where human oversight is required is the consequences of introducing machines that in some way cause stress to a workforce. This is subjective and humans can understand and possibly correct for this. A human may be able to explain why an AI system has come to a conclusion or diagnosis in way that is useful to another human.

5.1.2 Structure of a business case

In this section we'll look at the structure of a business case for AI. A business case is a formal document or presentation that outlines the justification for investment or a project. It is designed to help decision-makers assess the potential benefits, costs and risks associated with the initiative. The process of writing and producing a business case means the team starts learning from experience with informed decision-making. A well-constructed business case typically includes several vital elements: an executive summary that provides a concise overview of the proposal; a clear statement of the problem or opportunity that the project addresses; a detailed analysis of the proposed solution, including objectives, scope and deliverables; a thorough financial analysis that presents cost estimates, funding requirements and expected return on investment (ROI); an assessment of risks and mitigation strategies; and an implementation plan that outlines the timeline, milestones and responsibilities. Additionally, it should include an evaluation of alternatives considered and the rationale for the recommended approach. By incorporating these elements, a business case ensures a comprehensive evaluation of the project's feasibility and strategic alignment with organisational goals.

5.1.2.1 AI or ML business cases

Creating a business case for an AI or ML project involves several unique considerations compared to a traditional business case. The considerations highlighted next show the unique aspects of AI/ML projects that need to be addressed in a business case to ensure clarity, manage expectations and provide a clear path to achieving the desired outcomes. An example of the traditional format of a business case is:

- introduction;
- management or executive summary;
- description of current state;
- options considered;
- option described:
  - analysis of costs and benefits;
  - impact assessment;
  - risk assessment;
  - recommendations;
- appendices/supporting information.

With traditional projects, the technical feasibility is more straightforward, focusing on standard software or hardware implementations without the need for extensive data science or model training. Note this is weighted towards a ML project. However, the hardware is vital because it includes sensors and actuators. AI and ML projects require a detailed assessment of technical feasibility, including data availability and quality, the complexity of algorithms, model training and deployment challenges. You need to consider the infrastructure requirements and the integration of AI/ML models with existing systems. Remember our ethical needs embraced by a human-centric ethical purpose.

If we extend our ML project to include complex human and agent interactions with an environment, our project takes on a new dimension. Being more specific, the compute power (number of calculations) and memory (data storage) can take decades to plan and build; for example, weather modelling.

In projects over the last few decades we have seen that the technical feasibility is more straightforward, focusing on standard software or hardware implementations without the need for extensive data science or model training. If we embrace learning from experience or learning from data, then these projects require a detailed assessment of technical feasibility, including data availability and quality, the complexity of algorithms, model training and deployment challenges. We need to consider the infrastructure requirements and the integration of AI/ML models with existing systems. AI and ML algorithms have been developed by the scientific method, so data need to be well prepared for the algorithms. The success of AI/ML projects heavily depends on the data used for training models, and also the test data used to verify and validate the training of the models.

We are used to risks being more predictable and manageable, related to project scope, budget and timeline, whereas we are now moving into projects that carry higher risks and uncertainties due to factors such as model accuracy, over-fitting, biases in data and the need for continuous model maintenance and updates. The business case must address these risks and propose mitigation strategies. For example, an autonomous car is safe and reliable on the well-engineered motorway, but struggles with the nuances of country roads. This is the key business case for ML operations (MLOps)/ AI to IT operations (AIOps) or model operations (ModelOps) approaches; it provides key mechanism across the lifecycle to apply the necessary processes to mitigate risk as well as have the best change of intended outcomes.

Stakeholder engagement is often more focused on the business and IT alignment without the need for specialised data science roles. With our ethical and regulatory requirements, AI and ML projects will attract more exposure to a broader range of stakeholders. Involvement from a diverse set of stakeholders is crucial. The business case should outline how each stakeholder group will contribute and benefit from the project. Of particular importance is that a diverse stakeholder base will play a key role understanding bias and the outcome of a project. These may be AI ethics officers, ethics boards, legal (intellectual property, copyright), procurement, and so on.

Depending on the type of project, maintenance is usually related to software updates, hardware maintenance and user support, which are more predictable and less intensive. Learning from data and learning from experience bring with it more work and challenges. Continuous monitoring and maintenance of AI systems are necessary to ensure they remain accurate and relevant. The business case should include plans for ongoing support, model retraining and performance monitoring – as well as the maintenance of hardware.

Up to the present day, pilot projects (quick look, see, maybe) may not be as critical, and projects can move directly to implementation if the requirements and solutions are well understood. Agile project management and the minimum viable product has more emphasis on learning from experience, and the pilot is more commonplace. With this in mind, AI/ML projects, favouring a learning from experience approach, often start with a pilot or proof-of-concept phase to validate the model before full-scale implementation. The business case should include plans for scaling the solution based on pilot results.

Ethical implications and regulatory compliance are more prominent in ML and AI projects. Issues such as data privacy, model transparency and ethical AI practices need to be addressed in the business case. For example, a human-centric ethical purpose is at the core of the EU AI Act. While regulatory compliance is still important from an old-school IT perspective, ethical considerations are less likely to be a central focus unless the project directly affects data privacy or security. It is important to acknowledge, here, that there is a plethora of competing standards and guidance emerging globally, including from various UK government departments; however, in the UK particularly there is no clear regulatory or legislative guidance yet. Chapter 2 is focused on this specifically and will help to determine what is fit for purpose. A diverse team with a domain expert will help this.

5.2 STAKEHOLDERS AND PROJECT MANAGEMENT

In this section we look at stakeholders in more detail and what approaches to project management we can use if our project includes learning from experience or learning from data. For example, stakeholders will play a crucial role in helping organisations to comply with the EU AI Act by providing guidance, resources and oversight. Industry experts and academic researchers will offer insights into best practices for AI development and deployment, ensuring that organisations are well-informed about the regulatory requirements and ethical considerations. Civil society organisations will advocate for transparency and accountability, helping to monitor compliance and raise awareness about potential risks. Policymakers and regulatory bodies will provide clear and actionable guidelines, as well as support mechanisms such as training programmes and compliance tools. Through collaboration and knowledge sharing, stakeholders will assist organisations in implementing robust governance frameworks, conducting thorough risk assessments and maintaining continuous oversight of AI systems, thereby ensuring adherence to regulations and law (e.g. the EU AI Act, NIST, etc.) and fostering a culture of responsible AI innovation.

5.2.1 Stakeholder identification and categorisation

In AI/ML, stakeholders encompass a diverse group of individuals, organisations and entities that have an interest in or are affected by the development, deployment and use of AI technologies. With the heightened concern over our environment, the climate and biodiversity, these stakeholders are growing in influence and making a significant impact on the future directions of society and resources. AI/ML can contribute to intergenerational equity and sustainability in many ways.

Identifying stakeholders is crucial for effective stakeholder management, to ensure balanced consideration of interests, ethical implementation and responsible development of AI systems. One way of identifying stakeholders is to create a stakeholder wheel; an example is given in Figure 5.1. Remember that safety and security are vital to all stakeholders and should be designed in and continually improved.

5.2.2 Stakeholder power and influence

In the context of AI projects, the concepts of power and influence in stakeholder management remain crucial, but they take on specific nuances due to the nature of AI technology and its applications.

Power comes from formal roles within the organisation that authorise decision-making and resource allocation. It directly affects project scope, resource distribution and formal approval processes.

Figure 5.1 An example of an AI stakeholder wheel

Role: provide legal advice, ensure compliance, and address ethical concerns.

Role: establish regulations, guidelines and policies for AI development and use.

Role: develop and maintain AI systems, adapt to changes in job roles due to AI.

Role: develop AI technologies, commercialise AI products and services, drive innovation.

Role: fund AI start-ups and projects, influence business strategy.

Role: conduct fundamental and applied research, provide expertise, advance AI knowledge.

Role: advocate for ethical AI, represent societal interests, raise awareness.

Role: utilise AI products and services, provide feedback, drive demand.

Influence derives from expertise, reputation and the ability to persuade and mobilise others. It indirectly shapes the project's technical direction, user acceptance and innovative approaches.

5.2.2.1 Stakeholder power in AI projects

In AI projects, power refers to the formal authority stakeholders possess to make strategic decisions, allocate resources and enforce project directions.

Stakeholders with power have the following characteristics:

- **Strategic decision-making:** they can define the scope, goals and direction of the AI project.

- **Resource allocation:** they control budgets, personnel and technology infrastructure necessary for the AI project.

- **Regulatory compliance:** they ensure the AI project complies with legal, ethical and organisational standards.

- **Approval authority:** they have the final say on project milestones, deliverables and deployments.

Examples of power stakeholders could be:

- Executives and senior management who decide on the investment and strategic alignment of the AI project with business goals.

- Project sponsors who approve the AI project's budget and resource requirements.

- Legal and compliance officers who ensure the AI project adheres to regulatory standards and ethical guidelines.

5.2.2.2 Stakeholder influence in AI projects

In AI projects, influence refers to the ability of stakeholders to shape perceptions, decisions and actions related to the project without formal authority.

Stakeholders with influence have the following characteristics:

- **Expert knowledge:** they often possess deep technical expertise in AI, data science or related fields.

- **Thought leadership:** they can shape the vision and direction of the project through their insights and innovative ideas.

- **Networking and collaboration:** they can rally support and foster collaboration across different teams and departments.

- **User advocacy:** they represent the interests and needs of end-users, ensuring the AI solutions are user-centric and effective.

Examples of influence stakeholders could be:

- Data scientists and AI researchers who provide critical insights into the feasibility and design of AI models.

- Technical leads and engineers who influence the technical architecture and implementation of the AI solution.

- Ethics and privacy advocates who ensure the AI project considers ethical implications and user privacy concerns.

- Product managers who align the AI project with market needs and customer expectations.

- Ethics boards/committees who align the AI project with the ethics of the organisation.

5.2.3 Managing power and influence in AI projects

5.2.3.1 Strategies for power stakeholder management

Engage power stakeholders in high-level strategic planning, resource discussions and compliance checks using:

- Regular updates that provide structured reports to senior management and executives on project progress and challenges.

- Clear documentation that ensures all decisions, resource allocations and compliance measures are well-documented and transparent.

- Strategic alignment that continuously aligns the AI project's goals with the overarching business objectives – maintain executive support.

5.2.3.2 Strategies for influence stakeholder management

Engage influential stakeholders by involving them in brainstorming sessions, technical reviews and user feedback processes using:

- collaborative workshops with brainstorming sessions to gather insights from technical experts and innovators;
- communication channels that establish open and frequent communication with influential stakeholders to incorporate their feedback;
- recognition and inclusion that acknowledges the contributions of influential stakeholders and includes them in key decision-making processes to leverage their expertise.

By understanding and appropriately managing both power and influence, AI project managers can navigate the complexities of AI development, ensuring that both the organisation's strategic and technical perspectives are balanced and the project progresses smoothly towards its goals.

5.2.4 Project management approaches

In this section we look at how AI projects can be approached. We concentrate on two styles, Agile and waterfall (Wikipedia Contributors, 2024p and 2024q), along with the possible choice of a hybrid approach. Remember that AI is about learning from experience and ML is about learning from data. In an AI project, we often encounter tasks involving data learning. Here, we will focus on an ML project to align with the specific stages of ML workflow. We will describe a project in terms of software and hardware, but note that management of stakeholders is vital in the definition of the functionality that we need in our project. Note also that this is not a project management course, and larger scale projects will more than likely involve a mix of project management styles.

Selecting the best project management approach depends on the specific characteristics of the AI project, including the clarity of requirements, the need for flexibility, the level of stakeholder involvement and regulatory considerations.

There are many methodologies within project management, but they broadly fit into three main categories:

- Agile;
- waterfall;
- hybrid.

5.2.4.1 Agile project management for AI projects

Agile project management is an iterative and flexible approach to managing projects, particularly in software development, that emphasises adaptability, collaboration and customer satisfaction. Rooted in the Agile Manifesto, it prioritises individuals and interactions over processes and tools, working software over comprehensive documentation, customer collaboration over contract negotiation and responding to change over following a fixed plan. Agile methods break projects into small, manageable units called sprints or iterations, typically lasting two to four weeks, allowing teams to regularly reassess and adjust project direction based on feedback and evolving requirements. This approach fosters continuous improvement, rapid delivery of value and a high degree of transparency and communication within the project team and with stakeholders.

Agile projects' key characteristics are:

- **Iterative development:** projects are broken down into small, manageable units called sprints or iterations, typically lasting 2–4 weeks.

- **Flexibility and adaptability:** Agile allows for changes and revisions throughout the project lifecycle, adapting to evolving requirements and feedback.

- **Collaborative and cross-functional teams:** Agile emphasises close collaboration among team members and stakeholders, with daily stand-up meetings and continuous communication.

- **Continuous delivery and improvement:** Agile focuses on delivering functional components early and frequently, allowing for ongoing testing, feedback and improvement.

- **Customer-centric:** Agile prioritises customer needs and involvement, ensuring that the final product aligns closely with user expectations and requirements.

Its key characteristics mean that an Agile project style is suitable for AI projects because it has:

- **Rapid prototyping and feedback:** ideal for AI projects where quick prototyping and iterative feedback are essential for refining algorithms and models.

- **Uncertain and evolving requirements:** suitable for AI projects with evolving requirements or where the final outcome is not fully defined at the outset.

- **Innovation and experimentation:** which are crucial in AI development.

As an individual or an organisation becomes more skilled and experienced in AI projects and their management, some of the flexibility can be refined and made more efficient because the experience gained allows the organisation to predict and plan with more certainty. Agile style project management is complimented by MLOps.

5.2.4.2 Waterfall project management for AI projects

Waterfall project management is a sequential step-by-step approach to managing projects, traditionally used in industries such as construction and manufacturing but also applied in software development. Typical distinct phases are requirement gathering, design, implementation, testing, deployment and maintenance. Each is completed before moving on to the next. The waterfall model emphasises thorough planning, documentation and a clear definition of requirements at the project's outset, ensuring that all aspects of the project are well understood and agreed upon before development begins. Once a phase is completed, it is typically not revisited, which makes this approach less flexible in accommodating changes. This rigidity can be beneficial for projects with well-defined goals and stable requirements where predictability and a structured progression are critical for success. Can you conduct an AI project using a waterfall methodology? Of course you can, as long as you understand the end-game before you kickstart the project. Unfortunately, many AI projects don't know the end point from the outset, so naturally lend themselves towards a more Agile approach to project management, which incorporates a certain amount of experimentation.

Waterfall projects' key characteristics are:

- **Sequential phases:** projects progress through a series of distinct phases (requirements, design, implementation, testing, deployment) in a step-by-step fashion.

- **Fixed requirements:** requirements are gathered and defined upfront, and changes are typically minimised after the project begins.

- **Documentation:** thorough documentation at each phase is emphasised, ensuring a clear record of project decisions and processes.

- **Predictability and control:** projects follow a structured approach with clear milestones and deliverables, making it easier to manage and control the project timeline and budget.

Its key characteristics mean that a waterfall project style is suitable for AI projects because it has:

- **Well-defined requirements:** suitable for AI projects with clearly defined requirements and objectives, where changes are unlikely.

- **Regulatory and compliance needs:** ideal for projects that require extensive documentation and adherence to regulatory standards.

- **Large-scale infrastructure:** appropriate for large-scale AI infrastructure projects where a sequential approach ensures systematic progress and risk management.

5.2.4.3 Hybrid Agile/waterfall project management for AI projects
Hybrid Agile and waterfall project management is an approach that combines the structured, step-by-step processes of the waterfall model with the iterative, flexible practices of Agile methodologies. This hybrid model leverages the strengths of both approaches, using waterfall's clear, upfront planning and well-defined stages for parts of the project that are stable and predictable, while applying Agile's adaptability and iterative development cycles to areas that require flexibility and rapid feedback. This dual approach allows organisations to maintain control and predictability in project management while also being responsive to changing requirements and continuous improvement. By integrating the two methodologies, hybrid project management aims to maximise efficiency, mitigate risks and deliver high-quality results that align with both initial project scopes and evolving client needs.

Hybrid projects' key characteristics are:

- **Combines Agile and waterfall:** integrates elements of both Agile and waterfall methodologies, leveraging the strengths of each approach.

- **Flexibility and structure:** allows for flexibility in certain phases of the project while maintaining structured, sequential processes where necessary.

- **Customised approach:** tailors the project management approach to the specific needs and context of the project, balancing agility with predictability.

- **Incremental delivery with fixed milestones:** enables incremental delivery of project components while adhering to fixed milestones and deadlines.

Its key characteristics mean that a hybrid project style is suitable for AI projects because:

- **It deals with complex projects with mixed requirements:** suitable for AI projects that have both well-defined and evolving requirements, allowing different parts of the project to use different methodologies.

- **It has a balanced approach:** ideal for projects that need the flexibility of Agile for development and experimentation, coupled with the control and documentation of waterfall for implementation and compliance.

- **It includes stakeholder management:** effective in projects where different stakeholders require different levels of involvement and where phased delivery is necessary.

5.2.4.4 Selecting the appropriate project management approach for AI projects

The most appropriate project management style depends on the objectives of the protect and the scale of it.

- **Agile:** best for AI projects that require rapid iteration, constant feedback and adaptability. Suitable for experimental AI solutions, ML model development and projects with evolving user requirements.

- **Waterfall:** appropriate for AI projects with stable, well-defined requirements, such as regulatory compliance systems, large-scale data infrastructure and projects with strict timelines and budgets.

- **Hybrid:** optimal for AI projects that need both flexibility and structure, such as large enterprise AI implementations, projects with multiple phases that vary in complexity and when balancing innovation with compliance is necessary.

MLOps and ModelOps are frameworks that organisations have adopted to streamline the deployment, monitoring and maintenance of machine learning models in production.

5.3 RISK, COSTS AND GOVERNANCE

This section makes sense of what AI governance means and the need for it. Of course, governance, risks and costs are context specific and relate to what you and your organisation currently do. Some additional areas that need to be understood for AI elements you are working with or incorporating are also mentioned.

5.3.1 Governance activities in AI implementation

Effective governance in an AI implementation is crucial for ensuring compliance, managing risks and overseeing the lifecycle of AI models. The three primary areas of AI governance are:

- compliance to satisfy regulations, law and standards;
- risk management frameworks to proactively detect and mitigate risks; and
- lifecycle governance to manage, monitor and govern AI.

5.3.1.1 Compliance to satisfy regulations

Compliance ensures that AI systems are tested and adhere to existing laws, regulations and standards. This area of governance addresses the following aspects:

- **Regulatory adherence:**
  - Data protection – ensure AI systems comply with data protection laws such as GDPR, ensuring user data privacy and security.
  - Fairness and non-discrimination – adhere to laws that prevent bias and discrimination, ensuring AI systems provide fair outcomes across different demographic groups.
  - Transparency requirements – meet regulatory demands for transparency, making AI decision-making processes understandable and explainable to regulators and the public.

- **Auditability and accountability:**
  - Regular audits – implement regular audits of AI systems to verify compliance with regulatory standards (e.g. ISO/IEC 42001) and law (e.g. EU AI Act).
  - Traceability – maintain records of AI system development, data sources and decision-making processes to provide traceability in case of investigations or audits.
  - Safety – maintain records of AI system safety.

- **Ethical standards:**
  - Ethical guidelines – develop and follow ethical guidelines that align with industry standards and regulatory expectations.
  - Human oversight – ensure that human oversight is embedded in critical decision-making processes to maintain ethical standards and accountability.

5.3.1.2 Risks, costs and benefits analysis

Implementing an AI project or solution involves careful consideration of various risks, costs and benefits. Identifying and assessing these potential factors is crucial for successful implementation and alignment with the organisation's risk strategy. By identifying these risks, costs and benefits, and implementing appropriate mitigation strategies, organisations can align their AI projects with their overall risk strategy and ensure successful implementation. Some industries already have existing regulations, policies and standards for managing risk, which would need to be complied with (e.g. safety critical industries, nuclear, aviation, healthcare, etc.).

Proactive risk management involves identifying, assessing and mitigating risks associated with AI systems. This area of governance includes:

- **Risk identification and assessment:**
  - Risk frameworks – develop comprehensive risk frameworks to identify potential risks in AI systems, including technical, operational, ethical and legal risks.
  - Impact analysis – conduct impact analysis to understand the potential consequences of identified risks on stakeholders and the organisation.

- **Mitigation strategies:**
  - Bias mitigation – implement strategies to detect and mitigate biases in AI models, ensuring fairness and equity.
  - Robust security measures – strengthen security protocols to protect AI systems from cyber threats and data breaches.
  - Incident response plans – develop and maintain incident response plans to address and mitigate the effects of AI system failures or malfunctions.

- **Continuous monitoring:**
  - Real-time monitoring – use real-time monitoring tools to continuously assess the performance and behaviour of AI systems.
  - Feedback loops – establish feedback loops to gather input from users and stakeholders, allowing for timely identification and correction of issues.

We may find various types of risk that we'll need to manage, including:

- **Technical risks:**
  - System failures – AI systems may fail due to bugs, errors or unforeseen conditions.
  - Scalability issues – AI models might not scale well with increased data or usage.
  - Integration challenges – there may be difficulties in integrating AI with existing systems and processes.

- **Data risks:**
  - Data privacy and security – there might be risk of data breaches and misuse of sensitive information.
  - Data quality – poor-quality or biased data can lead to inaccurate or unfair AI outcomes.
  - Data availability – insufficient data can hinder the training and effectiveness of AI models.

- **Ethical and legal risks:**
  - Bias and discrimination – AI models may perpetuate or amplify biases present in the training data.
  - Regulatory compliance – compliance with laws and regulations, such as GDPR, must be ensured.
  - Transparency and accountability – there might be difficulty in explaining AI decisions and assigning responsibility.

- **Operational risks:**
  - Change management – resistance from employees and stakeholders to adopt AI solutions must be overcome.
  - Dependence on vendors – there could be over-reliance on third-party AI solutions and lack of in-house expertise.
  - Maintenance and monitoring – continuous monitoring and updating of AI models are both required to maintain performance.

Risk ownership is the assignment of responsibility to an individual or team within an organisation for managing a specific risk. The risk owner is accountable for identifying, assessing, monitoring and mitigating the assigned risk. Here are the key aspects of risk ownership:

- Responsibility and accountability – the risk owner is responsible for ensuring that the risk is appropriately managed according to the organisation's risk management policies and procedures, and accountability means the risk owner must report on the risk's status and any mitigation actions taken.
- Identification and assessment – the risk owner identifies the potential impact and likelihood of the risk and assesses the significance of the risk in the context of the organisation's objectives and operations.
- Mitigation and control – the risk owner develops and implements strategies to mitigate or control the risk, and this includes selecting appropriate risk mitigation measures such as avoidance, reduction, transfer or acceptance.
- Monitoring and reporting – the risk owner continuously monitors the risk and the effectiveness of the mitigation strategies and regularly reports on the risk status to senior management or relevant stakeholders, providing updates on any changes or emerging issues.
- Action plan development – the risk owner creates and maintains a risk action plan detailing specific steps to manage the risk, and this plan includes timelines, resources required and key performance indicators to track progress.

Risk ownership ensures that risks are managed proactively and systematically, with clear accountability and responsibility within the organisation. This structured approach helps in effectively addressing potential threats and opportunities. In particular, we can mitigate risks; various risk mitigation strategies can be applied to address a wide range of risks. Here are some key risk mitigation strategies:

- Risk avoidance – eliminate the risk entirely by not engaging in the activity or condition that introduces the risk.
- Risk reduction (mitigation) – reduce the likelihood or impact of the risk through various measures.
- Risk transfer – shift the risk to a third party, usually through insurance or outsourcing.
- Risk acceptance – acknowledge the risk and choose to accept it without taking any steps to mitigate it, often due to cost–benefit considerations.

- Risk sharing – distribute the risk among multiple parties, such as through partnerships or joint ventures.
- Risk prevention – implement measures to prevent the risk from occurring.
- Risk control – implement measures to control the impact of the risk if it occurs.
- Contingency planning – develop plans to manage and respond to the risk if it materialises.
- Diversification – spread the risk across multiple areas to minimise impact.
- Hedging – use financial instruments or other strategies to offset potential losses.
- Risk monitoring – continuously monitor the risk environment and the effectiveness of mitigation strategies.
- Training and education – educate and train employees and stakeholders on risk management practices and protocols.

With our better understanding of risks with AI, and managing and mitigating those risks, it's sensible to look at the costs of an AI project. They can be divided into three broad areas:

1. **Development and deployment costs:**
   - Software and hardware – investment in specialised software, tools and high-performance hardware.
   - Infrastructure – setting up the necessary IT infrastructure to support AI workloads.
   - Data acquisition – costs associated with collecting, purchasing or generating high-quality data.

2. **Human resources:**
   - Hiring experts – costs of hiring data scientists, AI engineers and other specialists.
   - Training – training existing staff to work with AI technologies and systems.

3. **Maintenance and upkeep:**
   - Ongoing support – regular updates, monitoring and troubleshooting of AI systems.
   - Operational costs – costs associated with running AI applications, such as cloud services.

With the broader understanding of the risks and costs of an AI project, here are some typical benefits:

- **Efficiency and productivity:**
  - Automation of tasks – AI can automate routine and repetitive tasks, freeing up human resources for more strategic activities.
  - Improved decision-making – AI can analyse large volumes of data quickly and provide insights to support decision-making.

- **Cost savings:**
  - Operational efficiency – AI can reduce operational costs through process optimisation and automation.
  - Resource optimisation – AI can provide better utilisation of resources through predictive analytics and intelligent resource management.
- **Enhanced customer experience:**
  - Personalisation – AI can provide personalised experiences and recommendations to customers.
  - 24/7 support – AI-powered chatbots and virtual assistants can offer continuous customer support.
- **Innovation and competitive advantage:**
  - New products and services – AI can enable the development of new products and services.
  - Market differentiation – early adoption of AI can provide a competitive edge in the market.

5.4 SUMMARY

This chapter explored how AI and ML provide opportunities for individuals, teams, organisations and society. It highlighted how AI enhances daily life through applications such as smartphones, smart devices and search engines. AI not only augments human capabilities but also fosters collaboration between humans and machines, driving improvements in productivity, decision-making and innovation.

The chapter delved into how AI can augment both humans and machines, introducing intelligent automation as a means to assist humans rather than replace them. It also presented the structure of a business case for AI projects, emphasising the importance of addressing technical feasibility, data strategies, risks, stakeholder engagement and ethical considerations. Unique challenges in AI projects, such as data quality and continuous model maintenance, make careful planning and pilot phases critical for success.

Stakeholders play a key role in AI projects, and the chapter outlined strategies for managing their power and influence. Various project management approaches – Agile, waterfall and hybrid – were explored, each suited to different project needs and complexities.

The chapter concluded with discussions on AI governance, emphasising compliance, risk management and lifecycle monitoring to ensure AI systems adhere to regulations and ethical standards. It also outlined potential risks, such as data privacy, scalability and bias, along with strategies for mitigating these risks. Finally, it highlighted the costs and benefits of AI, from development and maintenance to efficiency gains, cost savings and market differentiation, illustrating the value AI brings to organisations and society.

6   FUTURE DIRECTIONS AND AI CAREERS

In this chapter we look at how AI and ML will evolve from the immediate near future through to the distant future. In particular, how careers may evolve, and which ones are emerging in this exciting new fourth industrial revolution. Could you jump backwards 20, 30 or 40 years and foresee what you are doing today? Could you have predicted the technologies and systems we are using now? For some, the world we live in today is a disappointment as we were, arguably, promised an abundance of personal robots, flying cars and affordable spaceflight. We are still waiting for that to become commonplace; although, for what it's worth, we do have a couple of autonomous rovers on the planet Mars.

Forecasting where we will be in 2030, 2040 or even 2050 is equally as difficult, but we try based on current research and technology innovations that are in the pipeline. If you want to see different possible outcomes, Max Tegmark's book, *Life 3.0* (2018), is an intelligent and perceptive exploration of possible AI futures, and Ray Kurzweil's book, *The Singularity is Near* (2016), is an optimistic look at the future.

The are many benefits of AI. Social media frenzy, television programmes and movies all extol the success of possible futures of AI, but we need to be practical because chasing ivory-tower academic thought-leader experiments won't bring to life the benefits. As we move forward, we need to be explicit and careful about what AI and ML can achieve.

Paul R. Daugherty and H. James Wilson have written about their experience doing just that as management consultants in *Human + Machine: Reimagining Work in the Age of AI* (2018). In their book, they observe how AI is applied in many projects. They describe human-only systems, machine-only systems and human-plus-machine systems. They paint a picture of how the future may look, what roles humans and machines will play and what humans and machines will do together. There will be human-only, machine-only and human-plus-machine systems.

6.1 THE ROLES AND CAREER OPPORTUNITIES PRESENTED BY AI

In this section we will describe the human-only, machine-only and human-plus-machine systems, what roles these may lead to and what is already here.

6.1.1 Human-only systems

What do humans do well?

- leadership and goals;
- creativity;
- empathy;
- judgement.

In Chapter 1 we marvelled at what the human being is capable of. Machines are some way off this – decades or perhaps centuries off. Humans are good at the subjective viewpoint. We are creative; for instance, only a human, in this case Einstein, could have imagined a thought experiment of travelling on a light beam. It changed our understanding and the world as we know it today.

So, humans will be the creative powerhouse. We will let machines do the heavy lifting, leaving us to do the higher value work. Ambiguity is also a role for humans – what happens if the AI is stuck because of a contradiction? It is up to a human to step in and determine the outcome. We can think of many situations where human intervention might be needed (e.g. a confused dementia patient unable to answer questions, or an ethical 50/50 contradiction). A human who is confused could easily confuse an AI system; other humans need to be on hand to support that person.

Humans must also provide the leadership; we must set the goals and the ethical standards. These will form principles, values and rights. We will rely on these in law and, where the judgement is subjective and ambiguous, it will be humans who do this high value work.

6.1.2 Machine-only systems

What do machines do well?

- monotonous tasks;
- prediction;
- iteration;
- adaption.

These four points describe what machines do well, and this is easy to see with digital computers. We use them to do monotonous tasks such as lots of accurate calculations. Predicting and iterating are useful in digital computers for the solution of our ML problems – lots of calculations, repetitive iterations, finding a solution. Adaption is a trait of future machines, when machines will adapt to the needs of the day. Machines in this sense will also adapt to extreme environments; these could be deep in the ocean or on other planets where humans are exposed to extremes. More practically, preparing food in sterile conditions or operating on humans in an operating theatre could be better done by machines.

6.1.2.1 Narrow (weak) AI, AGI and strong AI

Weak or narrow AI is AI focused on a specific task. Popular ML of today, the capability we find on cloud digital services, is narrow AI, or narrow ML, focused on a specific task (e.g. supervised and unsupervised learning). Examples of narrow ML are SVMs, decision trees and KNN algorithms. They test hypotheses based on a specific task. The learning from experience is really learning from data (ML) on a specific, focused task.

AGI, introduced in Chapter 1, is a machine capable of learning any intellectual task that a human can. It is hypothetical and not required for AI and ML in general. This hypothetical type of AI is often the subject of speculation and sometimes feared by humans. We are nowhere near AGI, and it could be decades or even centuries before we are close to achieving it.

AGI can be taken one step further. Science fiction is fond of machines that are assumed to have consciousness; strong AI is AGI that is also conscious. Consciousness is complex and difficult. Some consider it to be the hardest problem in AI. This is covered in more detail later in the chapter.

AGI and strong, or conscious, AGI is not currently feasible and is not realistic in the foreseeable future. It is an active area of research, but not a requirement for AI or ML.

6.1.3 Human-plus-machine systems

Rather than thinking of either a machine or a human undertaking a task, surely the future is one with humans and machines in harmony? Or, as Stuart Russell emphasises, human-compatible AI – this concept is a proposal to solve the AI control problem. Russell's three principles are (Russell, 2019):

'1. The machine's only objective is to maximise the realisation of human preferences.

2. The machine is initially uncertain about what those preferences are.

3. The ultimate source of information about human preferences is human behaviour.'

Machines designed in this way will control the development of AI. Ideally, humans will move to higher value work or have more time to enjoy being human. These principles are new, and it's not yet clear how they will pan out; however, Stuart Russell suggests they are futureproof.

Practical examples of humans plus machines can be easily imagined. A person using an exoskeleton is able to undertake tasks they weren't previously capable of, perhaps allowing them to walk or lift heavy weights. We have search engines and large language models – academics can search literature in seconds, something that would have taken years, if not decades, 40 years ago. In medicine, brain interfaces and the study of the human brain is allowing doctors to better understand stroke recovery. Indeed, we may be able to test for consciousness in patients that we can't communicate with. The idea of humans and machines together gives us superhuman capabilities.

6.1.4 Roles and career opportunities in AI

Here we take a brief look at typical AI/ML roles. The list is not complete because the field of AI and ML is growing very quickly. However, it should give you a flavour of what we expect and currently see. We look at the roles and responsibilities of an ML engineer, data scientist, AI research scientist, computer vision engineer, NLP engineer, robotics engineer, AI ethics specialist and AI anthropologist.

An ML engineer will typically have the following role and responsibilities:

- **Developing algorithms:** design, build and implement ML models and algorithms.
- **Data processing:** clean, pre-process and organise data for use in models.
- **Model training and optimisation:** train models using relevant data sets and optimise them for better accuracy and performance.
- **Deployment and maintenance:** deploy ML models into production and maintain them over time.
- **Collaboration:** work closely with data scientists and software engineers to integrate models into applications.

A data scientist will typically have the following role and responsibilities:

- **Data analysis:** extract insights from structured and unstructured data using statistical techniques.
- **Model building:** develop predictive models and ML algorithms.
- **Visualisation:** create visualisations to communicate findings to stakeholders.
- **Experimentation:** design and conduct experiments to test hypotheses and improve models.
- **Business solutions:** translate business problems into data science problems and provide actionable solutions.

An AI research scientist will typically have the following role and responsibilities:

- **Research:** conduct advanced research in AI and ML to develop new algorithms and techniques.
- **Publications:** publish findings in academic journals and present at conferences.
- **Prototyping:** develop prototypes to test new AI concepts.
- **Collaboration:** work with academic and industry partners to push the boundaries of AI technology.
- **Innovation:** stay updated with the latest advancements in AI and apply them to solve complex problems.

A computer vision engineer will typically have the following role and responsibilities:

- **Image processing:** develop algorithms for image and video analysis.
- **Feature extraction:** implement techniques for object detection, recognition and tracking.

- **Model training:** train deep learning models specifically for computer vision tasks.
- **Optimisation:** optimise models for performance on different hardware and platforms.
- **Application development:** integrate computer vision capabilities into real-world applications such as AR/VR, autonomous vehicles, etc.

An NLP engineer will typically have the following role and responsibilities:

- **Text analysis:** develop algorithms for processing and analysing large amounts of natural language data.
- **Language modelling:** build and train models for tasks such as text generation, translation and summarisation.
- **Speech recognition:** implement systems for converting speech to text, and vice versa.
- **Sentiment analysis:** develop systems to understand and categorise the sentiments expressed in text.
- **Integration:** work on integrating NLP capabilities into applications such as chatbots, virtual assistants and search engines.

A robotics engineer will typically have the following role and responsibilities:

- **Design and development:** design, build and test robotic systems and components.
- **Programming:** develop software and algorithms for robot control and automation.
- **System integration:** integrate sensors, actuators and other hardware with robotic systems.
- **Testing and validation:** conduct tests to ensure reliability and performance of robots.
- **Research:** stay updated with the latest in robotics technology and apply it to innovate and improve existing systems.

An AI ethics specialist will typically have the following role and responsibilities:

- **Ethical analysis:** analyse the ethical implications of AI systems and their potential impact on society.
- **Policy development:** develop guidelines and policies to ensure ethical AI practices within the organisation.
- **Compliance:** ensure compliance with relevant laws, regulations and ethical standards.
- **Training and education:** educate employees and stakeholders on ethical AI practices and considerations.
- **Advisory:** provide advice on ethical dilemmas related to AI development and deployment.

An AI anthropologist will typically have the following role and responsibilities:

- **Human-centric research:** study the interaction between humans and AI systems, focusing on cultural and social impacts.
- **Ethnographic studies:** conduct ethnographic research to understand how AI is integrated into various communities and cultures.
- **User experience:** analyse and improve user experience by understanding human behaviour and cultural contexts.
- **Interdisciplinary collaboration:** work with engineers, designers and ethicists to ensure AI systems are culturally sensitive and user-friendly.
- **Impact assessment:** assess the long-term societal impacts of AI technologies.

6.2 AI IN THE REAL WORLD

AI tools and services are now part of the real world. First, we need to make sense of what this means in an organisation before looking at examples of AI. We need to consider a number of human–AI interactions:

- human in the loop (HITL);
- anthropology;
- ergonomics;
- human-centred AI (HCAI):
  - ethical AI design;
  - collaborative intelligence;
  - transparency and explainability;
  - inclusive design;
- modelling human behaviour in interactions with AI.

6.2.1 Human in the loop

Keeping HITL refers to the integration of human oversight and decision-making into AI systems. It ensures that humans remain actively involved in critical stages of AI processes, such as decision-making, monitoring and intervention (see Table 6.1).

Table 6.1 Human in the loop importance and example

Importance	Example
• Accountability – ensures that there is human accountability for decisions made by AI systems, especially in high-stakes or ethical scenarios. • Error correction – allows for human intervention to correct or override AI decisions that may be flawed or biased. • Trust and acceptance – enhances trust in AI systems by providing transparency and allowing human judgement in complex situations.	In healthcare, AI may assist in diagnosing diseases, but the final decision and treatment plan are reviewed and confirmed by a human doctor to ensure accuracy and consider nuanced patient factors.

6.2.2 Anthropology

Anthropology in the context of AI refers to the study of human behaviour, cultures and social interactions to inform the design and implementation of AI systems. It focuses on understanding how humans interact with technology and how AI can be designed to fit into various cultural and social contexts (see Table 6.2).

Table 6.2 Anthropology importance and example

Importance	Example
• User-centred design (UCD) – helps create AI systems that are more intuitive, culturally sensitive and user-friendly by understanding the diverse needs and behaviours of users. • Social impact – assesses the broader social implications of AI deployment, ensuring that AI systems enhance rather than disrupt social structures and relationships. • Ethical considerations – informs the ethical design of AI systems by considering cultural values and societal norms.	In designing a chatbot for customer service, anthropologists might study how different cultures communicate and interact with technology to ensure the chatbot's language and interaction style are appropriate and effective for a global audience.

6.2.3 Ergonomics

Ergonomics, also known as human factors engineering, involves designing AI systems and interfaces that optimise human wellbeing and overall system performance. It focuses on creating environments and tools that fit human physical and cognitive abilities (see Table 6.3).

Table 6.3 Ergonomics importance and example

Importance	Example
• Usability – ensures that AI systems are easy to use and reduce the cognitive load on users, enhancing productivity and reducing errors. • Comfort and safety – designs AI interfaces and environments that are comfortable and safe for users, minimising physical strain and health risks. • Efficiency – enhances efficiency by creating AI systems that users can interact with smoothly and intuitively.	In manufacturing, ergonomically designed AI interfaces for operating machinery can reduce the risk of repetitive strain injuries and make it easier for workers to interact with complex systems, thereby increasing efficiency and reducing errors.

6.2.4 Human-centred AI

User-centred design in HCAI involves designing AI systems with a deep understanding of the end-users' needs, preferences and contexts. This approach incorporates user feedback throughout the development process to create more intuitive and effective AI solutions (see Table 6.4).

Table 6.4 Human-centred AI importance and example

Importance	Example
• Enhanced usability – AI systems that are easier to use and understand, leading to higher adoption rates and user satisfaction. • Tailored experiences – personalised interactions and functionalities that cater to individual user needs and preferences. • Reduced errors – by considering user behaviour and cognitive load, systems are less prone to user errors.	In developing a personal health app, UCD ensures the app's interface is intuitive, provides relevant health insights and supports users in managing their health effectively.

HCAI is an approach to designing and deploying AI systems with a primary focus on enhancing human capabilities, ensuring meaningful user control and promoting human values. HCAI aims to create AI technologies that are not only technically effective but also socially and ethically responsible.

The key principles of HCAI are:

- User empowerment – AI systems should empower users by augmenting their abilities and supporting their goals rather than replacing human roles, and the aim is to create tools that enhance human decision-making, creativity and problem-solving.

- Transparency and explainability – AI systems should be transparent about how they operate and make decisions, for example XAI techniques are often employed to ensure that users can understand, trust and appropriately manage the AI's outputs.

- Ethical considerations – HCAI emphasises the ethical implications of AI, including fairness, accountability and mitigation of biases. It seeks to design AI systems that respect user privacy, ensure data security and promote equity.

- Usability and UX – the design of AI systems should prioritise usability and UX, ensuring that they are intuitive, accessible and responsive to user needs. This includes iterative testing and refinement based on user feedback.

- Collaboration and interaction – HCAI focuses on fostering effective collaboration between humans and AI systems. This encourages designs that facilitate smooth interaction, where AI systems act as cooperative partners rather than autonomous agents.

What HCAI does:

- Enhances human abilities – by providing tools that support human tasks, HCAI systems help users to achieve better outcomes, for instance AI can assist doctors in diagnosing diseases, support artists in creating new works or aid workers in optimising productivity.

- Improves decision-making – HCAI systems provide users with insights, recommendations and data analyses that support informed decision-making. These systems are designed to present information in a way that is easily understandable and actionable.

- Promotes trust and reliability – through transparency and explainability, HCAI fosters user trust in AI systems, for example users are more likely to rely on AI tools if they understand how decisions are made and can see the rationale behind AI-driven recommendations.

- Supports ethical use – HCAI ensures that AI technologies are developed and used responsibly, and incorporate ethical guidelines to avoid harm, ensure fairness and respect user rights, thereby contributing to public good and trust in AI technologies.

- Facilitates human–AI collaboration – by designing AI systems that work seamlessly with humans, HCAI encourages collaborative problem-solving and innovation. This includes interfaces and interaction models that allow users to effectively guide and control AI processes.

Applications of HCAI are:

- Healthcare – AI systems that assist clinicians by providing diagnostic suggestions, treatment recommendations and patient monitoring while ensuring the clinicians retain control over final decisions.
- Education – personalised learning systems that adapt to individual student needs and provide tailored educational experiences while supporting teachers in their instructional roles.
- Workplace productivity – AI tools that help workers manage tasks, optimise workflows and improve efficiency without overshadowing human judgement and expertise.
- Consumer technology – user-friendly AI in smartphones, smart homes and other consumer devices that enhance daily life by providing useful and easily controllable features.

HCAI represents a shift from viewing AI as a replacement for human labour to seeing it as a partner that enhances human capabilities. By prioritising user needs, ethical considerations and collaborative interaction, HCAI aims to create AI systems that are not only powerful and efficient but also trustworthy, transparent and aligned with human values.

6.2.5 Modelling human behaviour in interactions with AI

Several specialised models and frameworks have been developed for modelling human behaviour in interactions with AI. These models draw from psychology, human–computer interaction (HCI), cognitive science and behavioural economics. Here are some prominent ones:

- **Human–AI interaction models**
  - **Computational cognitive models:** simulate human thought processes to predict how users will interact with AI systems. Examples include ACT-R (adaptive control of thought – rational) and SOAR (state, operator and result).
  - **User modelling:** focuses on creating representations of user preferences, skills and behaviours to personalise interactions. Techniques include demographic models, psychographic models and behaviouristic models.
- **Behavioural economics and decision-making models**
  - **Prospect theory:** describes how people make decisions based on perceived gains and losses rather than final outcomes, which can be used to design AI systems that align with user expectations.
  - **Nudge theory:** proposes subtle changes in the environment or context to influence user behaviour in a predictable way without restricting options.

- HCI models

  - **The model human processor (MHP):** a framework that describes human interaction with systems in terms of perceptual, cognitive and motor processors.

  - **GOMS model (goals, operators, methods and selection rules):** analyses the user's cognitive structure and processes in the context of HCI tasks.

  - **Norman's interaction cycle:** describes the stages of interaction between users and systems, focusing on how users form goals, execute actions and interpret feedback.

- Trust and transparency models

  - **Trust in automation:** examines factors that influence user trust in AI systems, including transparency, reliability and user experience.

  - **XAI models:** focus on making AI decisions understandable to users to increase trust and facilitate better interaction.

- Affective computing models

  - **Emotion recognition and response:** systems designed to detect and respond to human emotions, enhancing user experience and engagement. Techniques include facial expression analysis, voice tone analysis and physiological signal detection.

- Social and collaborative interaction models

  - **Social presence theory:** investigates the degree to which a person feels 'present' in an interaction, which is critical for designing AI systems that engage users socially.

  - **Distributed cognition:** studies how cognitive processes are shared across individuals and artefacts in an environment, relevant for collaborative AI systems.

- Learning and adaptation models

  - **Reinforcement learning for personalised systems:** models user interactions to continuously improve recommendations and responses based on user feedback and behaviour.

  - **Interactive ML:** involves users in the training process of the AI, where user feedback helps to refine the model's predictions and behaviours.

- Ethical and reponsible AI frameworks

  - **Fairness, accountability and transparency in ML (FAT/ML):** frameworks that guide the design of AI systems to ensure ethical interaction, mitigating biases and ensuring accountability.

  - **UCD:** emphasises designing AI systems with a deep understanding of the end-users' needs, behaviours and contexts.

- Usability and UX models

  - **HCAI:** focuses on creating AI systems that enhance human capabilities and ensure meaningful user control.

  - **Technology acceptance model (TAM):** explains how users come to accept and use technology, considering perceived ease of use and perceived usefulness.

These models and frameworks are used to design, evaluate and refine AI systems to ensure they meet human needs, foster positive user experiences and facilitate effective and ethical human–AI interactions.

6.2.6 Real-world AI applications

In this section we look at the following broad applications of AI in the real world:

- marketing;
- healthcare;
- finance;
- transportation and logistics;
- education;
- manufacturing;
- entertainment;
- IT.

There are a multitude of examples on the internet of this rapidly expanding real-world application of learning from experience.

6.2.6.1 Marketing

AI is revolutionising marketing by enabling more personalised, efficient and effective strategies. Table 6.5 shows some ways that AI is used in marketing in additional detail and AI application.

Table 6.5 Uses of AI in marketing

Marketing area	Example
Personalised recommendations	E-commerce platforms – AI analyses user behaviour, purchase history and preferences to provide personalised product recommendations. This increases customer engagement and sales. For example, Amazon and Netflix use AI algorithms to suggest products and content tailored to individual users.
Customer segmentation	Targeted campaigns – AI helps marketers segment their audience based on various criteria such as demographics, behaviour and purchase history. This allows for more targeted and effective marketing campaigns. Tools such as Salesforce Einstein use AI to create detailed customer segments.

(Continued)

Table 6.5 (Continued)

Marketing area	Example
Chatbots and virtual assistants	Customer support – AI-powered chatbots handle customer inquiries in real time, providing instant support and information. This improves customer satisfaction and frees up human agents for more complex tasks. Examples include chatbots on websites and social media platforms such as Facebook Messenger.
Predictive analytics	Sales forecasting – AI analyses historical data to predict future trends, helping businesses make informed decisions about inventory, marketing spend and sales strategies. This predictive capability can optimise marketing efforts and improve ROI.
	Lead scoring – AI evaluates leads based on their likelihood to convert, allowing sales teams to prioritise high-potential prospects. Tools such as HubSpot and Marketo use AI for lead scoring.
Content creation	Automated writing – AI can generate content for blogs, social media and email marketing campaigns. This includes personalised emails and social media posts tailored to specific audiences. Tools such as Copy.ai and Jasper use AI to create engaging content quickly.
	Dynamic content – AI personalises website content in real time based on user behaviour and preferences. For instance, a user visiting an e-commerce site might see personalised product recommendations based on their browsing history.
Ad targeting and optimisation	Programmatic advertising – AI automates the buying of ads, targeting specific audiences more accurately and efficiently than traditional methods. Platforms such as Google Ads and Facebook Ads use AI to optimise ad placements and budgets in real time.
	Ad personalisation – AI creates personalised ad experiences by tailoring content and messaging to individual users based on their behaviour and preferences. This improves the relevance and effectiveness of advertising campaigns.
Sentiment analysis	Social media monitoring – AI analyses social media posts, reviews and comments to gauge public sentiment about a brand or product. This helps marketers understand customer opinions and address issues proactively. Tools such as Brandwatch and Hootsuite use AI for sentiment analysis.

(Continued)

Table 6.5 (Continued)

Marketing area	Example
Visual recognition	Image and video analysis – AI analyses images and videos to extract insights about customer preferences and behaviours. For example, AI can identify products in user-generated content on social media, helping brands understand which products are popular and how they are being used.
Email marketing	Personalised emails – AI crafts personalised email content based on user behaviour and preferences, increasing open rates and engagement. It also optimises send times and subject lines for maximum impact. Tools such as Mailchimp and SendGrid use AI to enhance email marketing campaigns.
Voice search optimisation	Search engine optimisation and content strategy – with the rise of voice-activated assistants such as Siri, Alexa and Google Assistant, AI helps to optimise content for voice search. This involves using NLP to understand and predict the questions users might ask.

By leveraging AI, marketers can enhance their strategies, improve customer engagement and achieve better results with more efficiency.

6.2.6.2 Healthcare

Healthcare is driven by ethical principles, and at its core is the scientific method. There appears to be a natural fit for AI, and it is about working out what is best for humans. Areas where AI is used in healthcare are set out in Table 6.6.

Table 6.6 Uses of AI in healthcare

Healthcare area	Example
Medical imaging and diagnostics	Radiology – AI algorithms, such as CNNs, are used to analyse medical images (X-rays, CT scans, MRIs) to detect abnormalities such as tumours, fractures and infections. For instance, AI tools can assist radiologists in identifying early signs of cancer with higher accuracy and speed.
	Pathology – digital pathology uses AI to analyse tissue samples for diagnosing diseases such as cancer. AI can identify patterns and anomalies in histopathological images, aiding pathologists in making quicker and more accurate diagnoses.

(Continued)

Table 6.6 (Continued)

Healthcare area	Example
Predictive analytics and risk assessment	Patient risk stratification – AI models analyse electronic health records to predict which patients are at higher risk of developing chronic conditions like diabetes or heart disease. These predictions help in early intervention and personalised treatment plans.
	Hospital readmission prediction – AI systems can predict the likelihood of a patient being readmitted to the hospital after discharge. This allows healthcare providers to implement targeted follow-up care and reduce readmission rates.
Personalised medicine	Genomic analysis – AI algorithms process genomic data to identify genetic mutations and variations associated with specific diseases. This information helps in developing personalised treatment plans based on a patient's genetic profile, particularly in oncology for targeted cancer therapies.
	Drug discovery – AI accelerates the drug discovery process by predicting how different compounds will interact with target proteins. AI models can screen millions of compounds to identify potential drug candidates, significantly reducing the time and cost involved in drug development.
Robotics and automation	Surgical robots – AI-powered robotic systems assist surgeons in performing precise and minimally invasive surgeries. For instance, the da Vinci Surgical System uses AI to enhance the surgeon's dexterity and precision during complex procedures.
	Automated pharmacy – AI-driven robots in pharmacies can accurately dispense medications, reducing errors and freeing up pharmacists to focus on patient care and counselling.
Virtual health assistants and chatbots	Patient interaction – AI chatbots provide patients with immediate responses to their health queries, assist in booking appointments and offer medication reminders. These virtual assistants improve patient engagement and access to healthcare information.
	Mental health support – AI-powered apps such as Woebot and Wysa provide mental health support through conversational AI, offering cognitive behavioural therapy techniques and emotional support to users.

(Continued)

Table 6.6 (Continued)

Healthcare area	Example
Remote monitoring and telehealth	Chronic disease management – wearable devices equipped with AI algorithms monitor vital signs and other health metrics of patients with chronic diseases. AI analyses the data in real time to detect anomalies and alert healthcare providers for timely intervention.
	Telemedicine – AI enhances telemedicine platforms by providing diagnostic support and decision-making tools to healthcare providers during virtual consultations. This ensures that patients receive accurate and timely care remotely.
Administrative efficiency	Clinical documentation – NLP tools transcribe and analyse clinical notes, streamlining documentation processes and reducing the administrative burden on healthcare providers.
	Resource allocation – AI optimises hospital operations by predicting patient admissions and discharges, improving staffing schedules and managing the supply chain of medical equipment and medications.
Early disease detection and screening	Cancer screening – AI algorithms analyse mammograms, colonoscopies and other screening tests to detect early signs of cancer. These tools can identify minute changes that may be missed by human eyes, leading to earlier diagnosis and treatment.
	Infectious disease outbreak prediction – AI models predict the spread of infectious diseases such as influenza or COVID-19 by analysing various data sources, including social media trends, travel patterns and historical outbreak data. This helps in planning and implementing preventive measures.

6.2.6.3 Finance

Finance is driven by ethical principles, and at its core is regulation. There appears to be a natural fit for AI. Areas where AI is used in finance are set out in Table 6.7.

Table 6.7 Uses of AI in finance

Financial area	Example
Fraud detection and prevention: real-time fraud detection	AI algorithms analyse transaction patterns and detect anomalies in real time, identifying potentially fraudulent activities and flagging them for further investigation. For instance, AI systems used by banks can immediately freeze accounts when suspicious activities are detected, protecting customers from fraud.

(Continued)

Table 6.7 (Continued)

Financial area	Example
Credit scoring and risk assessment: credit risk analysis	AI models evaluate a wider range of data points beyond traditional credit scores to assess the creditworthiness of individuals and businesses. This includes analysing social media activity, transaction history and other non-traditional data sources, leading to more accurate and inclusive credit assessments.
Algorithmic trading: high-frequency trading	AI algorithms execute trades at high speeds and volumes, capitalising on market opportunities within fractions of a second. These algorithms analyse market data, identify trends and make trading decisions without human intervention, enhancing profitability and efficiency.
Customer service and support: chatbots and virtual assistants	AI-powered chatbots provide 24/7 customer support, handling routine inquiries such as balance checks, transaction queries and account information. These virtual assistants improve customer service by providing quick and accurate responses, freeing up human agents to handle more complex issues.
Personalised financial advice: robot-advisors	AI-driven robot-advisors offer personalised investment advice based on individual financial goals, risk tolerance and market conditions. They automatically rebalance portfolios and provide insights, making financial planning accessible to a broader audience.
Risk management: predictive analytics	AI systems use predictive analytics to forecast market trends, asset prices and economic conditions. Financial institutions use these insights to manage risks more effectively and make informed investment decisions.
Regulatory compliance: regulatory technology solutions	AI helps financial institutions comply with regulatory requirements by automating the monitoring and reporting of transactions, detecting compliance breaches and ensuring adherence to anti-money laundering and Know Your Customer regulations.
Operational efficiency: process automation	AI automates routine back-office tasks such as data entry, reconciliation and report generation. This reduces human error, speeds up processes and lowers operational costs.
Investment research: sentiment analysis	AI analyses news articles, social media and other textual data to gauge market sentiment and predict stock movements. This helps investors and analysts to make more informed decisions based on real-time market sentiment analysis.
Portfolio management: AI-driven portfolio optimisation	AI algorithms optimise investment portfolios by continuously analysing market data and adjusting asset allocations to maximise returns and minimise risks based on predefined investment strategies.

6.2.6.4 Transportation and logistics

Transportation and logistics are driven by the need to organise and manage complexity. There appears to be a natural fit for AI. Areas where AI is used in transportation and logistics are set out in Table 6.8.

Table 6.8 Uses of AI in transportation and logistics

Transportation and logistics area	Example
Autonomous vehicles	Companies such as Tesla, Waymo and Uber use AI algorithms for navigation, obstacle detection and decision-making to enable autonomous driving. These vehicles utilise sensors and ML to perceive their environment and drive safely.
Traffic management and optimisation	Smart traffic lights – AI systems adjust traffic light timings in real time based on traffic flow data to reduce congestion and improve traffic flow efficiency. Cities such as Los Angeles and London have implemented such systems.
	Traffic prediction – AI models predict traffic patterns and provide real-time traffic updates to drivers through navigation apps such as Google Maps and Waze, helping to avoid congested routes and reducing travel times.
Predictive maintenance	Vehicle maintenance – AI analyses data from sensors on vehicles to predict maintenance needs before failures occur. This helps in scheduling timely maintenance, reducing downtime and extending the lifespan of vehicles.
	Infrastructure maintenance – AI systems monitor infrastructure such as bridges and roads, identifying wear and tear or potential failures, allowing for proactive maintenance and repairs.
Public transportation systems	Dynamic scheduling – AI optimises bus and train schedules based on passenger demand and traffic conditions, ensuring efficient and timely public transportation services. Cities such as Singapore and Amsterdam have implemented such systems.
	Passenger flow management – AI monitors and predicts passenger flow in public transport hubs, helping to manage crowd control and improve the passenger experience during peak times.

6.2.6.5 Education

Education is driven by ethical principles, and at its core is our quest for knowledge – we may even be bold enough to say that it is the epitome of learning from experience. There appears to be a natural fit for AI. Areas where AI is used in education are set out in Table 6.9.

Table 6.9 Uses of AI in education

Education area	Example
Personalised learning	Adaptive learning platforms – platforms such as Khan Academy and Coursera use AI to create personalised learning paths for students based on their strengths, weaknesses and learning pace. These systems adjust content and exercises to match individual learning needs.
	Intelligent tutoring systems – AI-powered tutoring systems such as Carnegie Learning provide real-time feedback and personalised tutoring to students, helping them to understand complex concepts and improve their skills.
Automated grading and assessment	Automated essay scoring – AI tools such as Grammarly and Turnitin's Revision Assistant use NLP to evaluate and grade written assignments, providing instant feedback on grammar, structure and content.
	Multiple-choice and short-answer grading – systems such as Gradescope automate the grading of multiple-choice and short-answer tests, significantly reducing the time educators spend on grading.
Virtual assistants and chatbots	24/7 student support – AI chatbots such as IBM's Watson Education provide students with instant answers to their questions, assist with administrative tasks and offer study tips, enhancing student support services.
	Virtual teaching assistants – AI teaching assistants such as Georgia Tech's 'Jill Watson' help to answer student queries and manage routine tasks, allowing human instructors to focus on more complex issues.
Content creation and curation	Platforms such as Smart Sparrow recommend additional resources and aim to create a more personalised, engaging and effective learning experience, ensuring that each student receives the support they need to succeed.

(Continued)

Table 6.9 (Continued)

Education area	Example
Learning analytics and insights	Student performance tracking – AI-driven analytics platforms such as Brightspace Insights track student performance and engagement, providing educators with actionable insights to improve teaching strategies and student outcomes.
	Predictive analytics – AI systems predict student success and identify at-risk students by analysing data on attendance, participation and grades, allowing for early interventions and support.
Language learning	AI language tutors – apps such as Duolingo use AI to provide personalised language lessons, adapt to user proficiency levels and offer instant feedback on pronunciation and grammar.
	Real-time translation – AI-powered translation tools such as Google Translate offer real-time translation and language practice, supporting multilingual education and communication.
Enhanced classroom experiences	Interactive learning environments – AI-powered tools such as Classcraft create interactive and gamified learning experiences, engaging students and enhancing classroom participation.
	AR and VR – AI integrates with AR and VR to create immersive educational experiences, such as virtual field trips and interactive simulations, enriching the learning process.
Administrative efficiency	Automated administrative tasks – AI systems streamline administrative processes such as enrolment, scheduling and attendance tracking, freeing up time for educators and administrators to focus on teaching and student support.
	Resource management – AI optimises resource allocation, such as classroom space and equipment usage, ensuring efficient use of educational resources.

6.2.6.6 Manufacturing

Manufacturing is driven by ethical principles, and at its core is the scientific method. There appears to be a natural fit for AI. Areas where AI is used in manufacturing are set out in Table 6.10.

Table 6.10 Uses of AI in manufacturing

Manufacturing area	Example
Predictive maintenance	Equipment monitoring – AI systems analyse data from sensors on machinery to predict when equipment is likely to fail, allowing for maintenance to be performed just in time to prevent unexpected breakdowns. Companies such as Siemens use AI-driven predictive maintenance to minimise downtime and extend equipment life.
Quality control and inspection	Automated visual inspection – AI-powered computer vision systems inspect products for defects during the manufacturing process, ensuring high quality and consistency. For instance, companies such as IBM use AI to detect defects in semiconductor manufacturing.
Supply chain optimisation	Demand forecasting – AI models predict future demand for products by analysing historical sales data, market trends and external factors. This helps manufacturers to optimise inventory levels and production schedules, as seen in companies such as Amazon and Walmart.
	Logistics optimisation – AI algorithms optimise routing and scheduling for logistics and supply chain operations, reducing transportation costs and improving delivery times.
Robotics and automation	Collaborative robots (cobots) – AI-powered cobots work alongside human workers to perform repetitive and physically demanding tasks, enhancing efficiency and safety. Universal Robots and FANUC are examples of companies using cobots in manufacturing.
	Automated guided vehicles (AGVs) – AI-driven AGVs transport materials and products within manufacturing facilities, optimising workflow and reducing manual handling.
Process optimisation	Production scheduling – AI systems optimise production schedules based on real-time data and constraints, improving throughput and reducing lead times. GE uses AI for optimising manufacturing processes in its factories.
	Energy management – AI monitors and optimises energy consumption in manufacturing plants, reducing costs and environmental impact. Siemens' AI systems help to manage energy use in industrial settings.

(Continued)

Table 6.10 (Continued)

Manufacturing area	Example
Product design and development	Generative design – AI algorithms generate optimised product designs based on specified constraints and requirements, often resulting in innovative and efficient designs. Autodesk's generative design tools are used in various manufacturing applications.
	Simulation and testing – AI-driven simulations test product designs and manufacturing processes virtually, identifying potential issues before physical prototypes are built.
Inventory management	Smart inventory systems – AI systems track inventory levels in real time, predict stock requirements and automatically reorder supplies to ensure optimal inventory levels. This reduces stockouts and overstock situations.
Safety and risk management	Worker safety monitoring – AI-powered systems monitor worker safety by analysing data from wearables and environmental sensors, detecting hazardous conditions and alerting workers and management.
	Risk assessment – AI analyses historical incident data and current conditions to assess risks and implement preventive measures, enhancing workplace safety.

6.2.6.7 Entertainment

Entertainment is driven by creativity and inspiration. It is a complex pursuit and draws significantly on technology. There appears to be a natural fit for AI. Areas where AI is used in entertainment are set out in Table 6.11.

Table 6.11 Uses of AI in entertainment

Entertainment area	Example
Content creation	AI-generated music – AI systems such as OpenAI's MuseNet and Amper Music create original music compositions, providing artists and producers with new tools for music creation.
	Script and story writing – AI tools such as ScriptBook analyse existing scripts and generate new story ideas, helping writers with plot development and character creation.

(Continued)

Table 6.11 (Continued)

Entertainment area	Example
Personalised recommendations	Streaming services – Netflix and Spotify use AI algorithms to analyse user preferences and behaviours, providing personalised recommendations for movies, TV shows and music.
	Content curation – Social media platforms such as YouTube and TikTok use AI to curate content feeds, ensuring users see content that matches their interests and engagement patterns.
Visual effects and animation	AI-enhanced animation – AI tools such as DeepMotion and Autodesk Maya use ML to automate and enhance animation processes, making character movements more realistic and reducing manual labour.
	Deepfake technology – AI is used to create deepfake videos, which can be applied in film production to recreate historical figures or de-age actors, as seen in movies such as *The Irishman*.
Interactive experiences	Video game development – AI is used to create more intelligent and adaptive non-playable characters in video games, improving gameplay and user engagement. Games such as *The Last of Us* and *Red Dead Redemption 2* utilise AI for advanced character behaviour.
	VR and AR – AI enhances VR and AR experiences by enabling more immersive and interactive environments. Applications such as Pokémon GO use AI to blend digital objects seamlessly with the real world.
Audience engagement and marketing	AI chatbots – AI-powered chatbots engage with fans on social media platforms, providing personalised interactions and answering questions about shows, movies or events. Disney uses AI chatbots to enhance customer engagement.
	Targeted advertising – AI analyses user data to create highly targeted advertising campaigns, increasing the effectiveness of marketing efforts. Platforms such as Facebook and Google Ads utilise AI for precise ad targeting.
Voice and speech recognition	Voice assistants – AI voice assistants such as Amazon Alexa and Google Assistant allow users to interact with entertainment systems using voice commands, controlling playback, searching for content and more.
	Speech synthesis – AI-powered speech synthesis, such as Adobe Voco, can create realistic voiceovers for movies, games and audiobooks, enhancing audio production quality.

(Continued)

Table 6.11 (Continued)

Entertainment area	Example
Real-time analytics and insights	Audience analytics – AI analyses real-time data on audience reactions and engagement during live events, helping producers and marketers to adjust strategies to maximise impact. Platforms such as Twitch use AI to monitor viewer engagement.
	Content performance optimisation – AI tools analyse how content performs across various platforms, providing insights that help creators to optimise their content for better reach and engagement.
Facial recognition and emotion analysis	Emotion-driven content – AI analyses viewers' facial expressions and emotions while watching content to provide feedback on engagement levels and emotional responses. These data can be used to tailor future content.
	Interactive shows – interactive TV shows and games, such as Netflix's *Bandersnatch*, use AI to offer viewers choices that affect the storyline, creating personalised viewing experiences.

6.2.6.8 Information technology

IT is universal to all areas of our lives. It is easy to fall into the trap of thinking that IT is the driver of AI in an organisation. This is almost intuitive because of the importance attached to ML. IT enables AI, and in particular ML. Some areas where AI fits into IT are set out in Table 6.12.

Table 6.12 Uses of AI in information technology

Information technology area	Example
IT infrastructure management	Automated system monitoring – AI-powered tools such as Dynatrace and Splunk monitor IT infrastructure in real time, detecting anomalies and potential issues before they cause significant problems. These tools use ML to analyse system performance and provide alerts for unusual behaviour.
	Predictive maintenance – AI algorithms predict hardware failures by analysing data from various sensors and logs, allowing for proactive maintenance and reducing downtime. For instance, IBM's Predictive Maintenance and Quality solution uses AI to foresee and prevent equipment failures.

(Continued)

Table 6.12 (Continued)

Information technology area	Example
Cybersecurity	Threat detection and response – AI-based cybersecurity solutions such as Darktrace and CrowdStrike use ML to detect unusual patterns and potential threats in real time, enabling quicker response to cyber-attacks.
	Fraud detection – financial institutions use AI to detect fraudulent activities by analysing transaction patterns and user behaviour. For example, PayPal uses AI to monitor transactions and flag suspicious activities.
IT service management	Automated ticket resolution – AI chatbots and virtual assistants such as ServiceNow Virtual Agent handle routine IT support requests, such as password resets or software troubleshooting, improving efficiency and reducing the workload on human agents.
	Intelligent incident management – AI tools prioritise and categorise IT service requests and incidents, ensuring that critical issues are addressed promptly. Tools such as Moogsoft use AI to reduce alert noise and correlate incidents, helping IT teams to focus on resolving major issues.
Data management and analytics	Data cleaning and preparation – AI tools such as Trifacta automate the process of cleaning and preparing data for analysis, saving time and reducing errors. These tools use ML to identify and correct data inconsistencies.
	Advanced analytics – AI-driven analytics platforms such as Tableau and Microsoft Power BI use ML to analyse large data sets, uncovering insights and trends that help organisations to make informed decisions.
Software development	Code generation and review – AI tools such as GitHub Copilot assist developers by suggesting code snippets and completing code segments, improving coding efficiency and reducing errors. AI-powered code review tools such as DeepCode analyse code for potential bugs and vulnerabilities.
	Automated testing – AI-based testing tools such as Testim and Applitools automate the software testing process, identifying bugs and performance issues more efficiently than manual testing.

(Continued)

Table 6.12 (Continued)

Information technology area	Example
Cloud computing optimisation	Resource allocation – AI algorithms optimise cloud resource allocation by predicting demand and adjusting resources, accordingly, reducing costs and improving performance. Tools such as Amazon Web Services Auto Scaling use AI to manage resources dynamically.
	Cost management – AI tools analyse cloud usage patterns and recommend cost-saving measures, such as rightsizing instances or shifting workloads to cheaper times. Solutions such as CloudHealth by VMware provide detailed insights and optimisation recommendations.
Network management	Network traffic analysis – AI tools monitor network traffic to detect congestion, optimise routing and improve overall network performance. Cisco's AI Network Analytics uses ML to provide insights into network operations and suggest optimisations.
	Anomaly detection – AI systems detect and alert administrators about unusual network activities that could indicate security breaches or performance issues. Solutions such as Juniper Networks' AI-driven wide area network management use ML to identify and address anomalies.
UX enhancement	Personalised user support – AI-driven virtual assistants provide personalised IT support to users, guiding them through troubleshooting steps based on their specific issues and preferences.
	Performance optimisation – AI tools analyse user interactions and system performance to suggest optimisations, ensuring smooth and efficient user experiences with applications and services.

6.3 FUTURE OF AI

Ray Kurzweil's book, *The Singularity Is Near: When Humans Transcend Biology* (2016), explores the concept of the technological singularity – a future point when AI surpasses human intelligence, leading to unprecedented technological growth and transformation.

Ray Kurzweil is a renowned American inventor, futurist and author known for his pioneering work in AI, technology and futurism. Often referred to as one of the leading visionaries of our time, Kurzweil has made significant contributions to fields such as OCR, text-to-speech synthesis and speech recognition technology. He is also a prominent advocate of the concept of the technological singularity – the idea that the rapid advancement of technology will lead to a future where artificial intelligence surpasses human intelligence, radically transforming society and human life. Kurzweil's predictions, many of which have been remarkably accurate, are based on his theory of the 'Law of Accelerating Returns', which suggests that the pace of technological progress is exponential. Through his books, including *The Singularity Is Near* (2016) and *How to Create a Mind* (2012), Kurzweil has influenced discussions on the future of humanity, often sparking both excitement and controversy with his bold forecasts about the intersection of technology and human evolution.

Kurzweil argues that this singularity will occur by the mid-21st century, driven by exponential advancements in fields such as computing, biotechnology and nanotechnology. He envisions a future where humans and machines merge, leading to radical changes in society, economy and even the nature of human existence itself. The book is both a detailed analysis of technological trends and a speculative look at the profound implications of a post-singularity world, where the boundaries between human and machine become increasingly blurred.

6.3.1 AI's impact on society and the future

AI is evolving rapidly. This rapid technological advancement comes with benefits and challenges at societal level. Our objective in this section is to make sense of these benefits and challenges and the impact on society. We'll do this by discussing the potential future of AI. Examples of the benefits include reducing human error through task automation, processing and analysing vast amounts of data for informed decisions (AI algorithms), AI-powered tools assisting in medical diagnosis and so on. Examples of the challenges are ethical concerns about algorithm bias and privacy, job loss, lack of creativity and empathy, security risks from hacking, socio-economic inequality, market volatility because of AI-driven trading algorithms and AI systems' rapid self-improvement, along with the potential future advancements and direction of AI (e.g. increased computing power, availability of more data, better algorithms, improved tools).

Predicting the exact advancements and direction of AI over the next 50 years is challenging, but we can outline potential trends and areas of development based on current research and emerging technologies. Some potential future advancements and directions for AI, along with their potential effects, are set out in Table 6.13.

The future of AI holds immense potential for transformative effects across all aspects of society. However, realising these advancements will require addressing ethical, technical and societal challenges to ensure that AI benefits all of humanity.

Table 6.13 The potential impact of AI in the future

AI advance	Potential impact
General artificial intelligence – development of AGI, which can perform any intellectual task that a human can do, is a significant long-term goal for AI researchers.	AGI could revolutionise industries by automating complex tasks, leading to unprecedented levels of efficiency and innovation. It could also pose significant ethical and control challenges.
Human–AI collaboration – enhanced collaboration between humans and AI systems, where AI augments human capabilities rather than replaces them.	AI assistants could become ubiquitous, helping with decision-making, creativity and productivity in everyday life and professional settings.
Advanced NLP – continued improvements in NLP, enabling AI to understand, generate and interact in human language more effectively and naturally.	Seamless communication with AI, including real-time language translation, more sophisticated virtual assistants and improved accessibility for people with disabilities.
Autonomy – advancements in AI around autonomy have the potential to transform multiple industries, particularly transportation, logistics, delivery services, healthcare and home automation.	Transformations in transportation, logistics and delivery services, along with advancements in healthcare (surgical robots) and home automation.
AI in healthcare systems – development of robust frameworks for AI ethics and governance to ensure responsible AI deployment.	Improved healthcare outcomes, more efficient healthcare delivery, early detection and treatment of diseases and reduced healthcare costs.
AI ethics and governance – integration of quantum computing with AI, enabling the solving of complex problems that are currently intractable for classical computers.	Ethical AI systems that align with human values, increased public trust in AI and mitigation of risks associated with AI misuse.
Quantum computing and AI – integration of quantum computing with AI, enabling the solving of complex problems that are currently intractable for classical computers.	Breakthroughs in cryptography, material science, complex system simulations and optimisation problems.
AI in climate and environmental science – use of AI to model and predict climate change, optimise renewable energy systems and manage natural resources.	Better climate change mitigation strategies, improved sustainability practices and enhanced environmental conservation efforts.

(Continued)

Table 6.13 (Continued)

AI advance	Potential impact
AR and VR – AI-enhanced AR and VR systems for immersive experiences in gaming, education, training and remote work.	New ways of learning and working, enhanced entertainment experiences and improved training simulations for various professions.
Emotional and social intelligence – development of AI systems that can understand and respond to human emotions and social cues.	More empathetic AI interactions, improved mental health support through AI and enhanced customer service experiences.
AI in education – personalised learning systems powered by AI that adapt to individual student needs and learning styles.	Improved educational outcomes, more inclusive education and lifelong learning opportunities.
AI in security and defence – advanced AI applications in cybersecurity to detect and counter threats, and in defence for strategic planning and autonomous defence systems.	Enhanced security measures, more effective defence strategies and ethical considerations in autonomous weapon systems.
Interdisciplinary AI research – increased collaboration between AI and other scientific fields, leading to interdisciplinary research and innovation.	New discoveries and advancements in fields such as biology, chemistry, physics and social sciences through the application of AI.
Energy efficiency and sustainability – AI-driven optimisation of energy usage in various sectors, from smart grids to manufacturing.	Reduction in energy consumption, increased use of renewable energy sources and overall improvements in sustainability practices.
AI for social good – AI applications aimed at solving global challenges such as poverty, hunger and inequality.	Significant contributions to achieving the United Nations' Sustainable Development Goals, leading to a more equitable and just world.

6.3.1.1 Potential problems of AI

The existential risks associated with the use of AI are serious and well-recognised by experts in the field. These risks primarily stem from the potential for AI to surpass human intelligence and operate in ways that could be harmful if not properly controlled or aligned with human values. These can be summarised as follows:

- **Misaligned objectives** – an AI system could develop goals that are not aligned with human values and interests and, if the AI has advanced capabilities, it might pursue these goals in ways that are detrimental to humanity.

- **Loss of control** – as AI systems become more autonomous and capable, there is a risk that humans might lose control over them and this could happen if the AI develops the ability to resist human intervention or if it becomes too complex for humans to understand and manage.

- **Unintended consequences** – even well-intentioned AI systems can have unintended consequences due to unforeseen interactions or emergent behaviours, and the complexity of AI systems can make it difficult to predict all possible outcomes.

- **Self-improvement and runaway AI** – a super-intelligent AI might enter a self-improvement loop, rapidly enhancing its own capabilities beyond human control, and this 'intelligence explosion' could lead to an AI that is vastly more powerful than any human oversight can manage.

- **Ethical and value alignment** – ensuring that AI systems operate according to human ethical standards and values is a significant challenge; the misalignment in ethical frameworks could lead to AI making decisions that are morally unacceptable.

- **Weaponisation** – AI technologies could be used to develop autonomous weapons and other military applications that operate without human oversight, leading to uncontrolled and potentially indiscriminate use of force.

- **Economic and social disruption** – the widespread implementation of AI could lead to significant economic and social disruptions, such as mass unemployment, inequality and social unrest and, while not necessarily existential, these disruptions could contribute to broader societal collapse. The internet and poor IT security has given AI the opportunity for widespread manipulative and criminal behaviour. These include the quick exploitation of security failures, phishing and SMS scams, fake videos, audio and text and social manipulation. In the latter example, this has led to alledged manipulation of voting in elections all over the world.

6.4 CONSCIOUSNESS AND ETHICAL AI

Human consciousness, sometimes referred to as sentience, is, in its simplest terms, having an awareness of an internal or external existence or having a mental state you are aware of being in. This can be compared to subconsciousness, which is that part of your mind that notices and remembers information when you are not actively trying to do so and can often influence your behaviour even though you do not realise it.

Our human experience is inherently subjective and the way we judge situations or actions varies from person to person. As a result, ethics becomes a complex area that anyone working with AI must carefully consider. Since ethics are shaped by individual perspectives, understanding it requires thoughtful consideration of everyone's unique experiences.

AI is steeped in the scientific method and uses a basis of objects to make sense of our random, non-linear and complicated world. To do this we use objective science. Objectivity strives to provide a clear and unbiased view of reality, minimising the influence of subjective factors and personal biases. In the simplest terms, objectivity is something we all agree on.

6.4.1 AI and consciousness

Some humans fear conscious machines, however unlikely they may be. Consciousness is a complex area at the cutting edge of AI research; our knowledge is growing, but we might never understand what consciousness actually is. Interfacing the human mind with machines could advance medical success or outcomes – examples that are currently being explored include stroke recovery.

The most influential modern physical theories of human consciousness have been proposed by numerous neuroscientists, physicists and philosophers over the last couple of centuries, and they seek to explain consciousness in terms of neural events occurring within the brain.

Nobody currently knows where consciousness actually comes from, but it's thought to be an emergent property that comes from neurons connected in a specific but complex way. We do understand how parts of our brain function, but much is still unknown. Interestingly, the study of organoids (e.g. small blobs of brain tissue grown in the laboratory) is allowing scientists to gain an understanding of how parts of the brain function.

In a broader physiological sense, our AI or more specific ML focuses on the cognitive, digital voice in our brain as being some sort of representation of consciousness – generative AI and large language models being our most up-to-date attempt. Our human whole body subjective experience draws on our brain, heart and stomach as well the interaction with our senses through the autonomous nervous system and vagus nerve. Where do you feel nervous? Describe love. As you can see, our AI and ML isn't anywhere close yet.

Machine consciousness or artificial consciousness is the domain of cognitive robotics and is primarily concerned with endowing an artificial entity with intelligent behaviour by providing it with a processing architecture. This will allow it to learn and reason about how to behave in response to complex goals in a complex world. The aim of the theory of artificial consciousness is to define what would have to be synthesised to develop artificial consciousness in an engineered artefact such as a robot.

Many organisations are pursuing developments in artificial consciousness, but comparing consciousness in functional machines to consciousness in functional humans is more difficult than expected and the topic has raised a debate around the risks of developing conscious machines.

Rather than sprinting into a rabbit warren of opinion, let's look at how we may, in general, describe consciousness, the state of being aware of and able to think about one's own existence, thoughts, and surroundings; or, as René Descartes (1596–1650) expressed it (Wikipedia Contributors, 2024b):

I think, therefore I am.

We can notice straight away that AI, an objective science, may be used to study our subjective experience. AI can certainly help us with understanding complex and random systems way beyond a human's capabilities. Anil Seth, a scientist who has devoted 20 years to the study of the human brain, describes that there is no requirement of intelligence for something to be conscious (Seth, 2021). His research emphasises that consciousness is about having subjective experiences and awareness, whereas intelligence is related to problem-solving abilities, learning and adaptability. This isn't good news for those who have the view that today's digital AI (ML really) will suddenly become conscious. Perhaps it will reassure those who fear AI; however, we simply don't know how AI will ultimately evolve, but its effect on humans and society as a whole must be a priority. Indeed, in 1972 Herbert Dreyfus predicted we would come to this end point in his book, *What Computers Can't Do*.

Francis Crick is more frank about what consciousness is in his book, *The Astonishing Hypothesis: The Scientific Search for the Soul* (1994):

> you, your joys and your sorrows, your memories and your ambitions, your sense of personal identity and free will, are in fact no more than the behaviours of a vast assembly of nerve cells and their associated molecules.

The hard question of consciousness, introduced in Section 1.1.1.6, makes us really understand the subjective nature of our feelings and that we have no scientific way to define why we have them. This is an exciting endeavour for researchers in many fields and, if we get it right, might lead to better understanding of being human.

No machine has achieved anything even close to consciousness – to date. Remember that we can't define it, yet alone simulate it. Even if we could, researchers think it will be as complex as quantum mechanics. It remains a research challenge, but some leading academics believe that it will be possible to generate consciousness in a machine (decades away). There are numerous risks if we do develop machine consciousness, and most leading academics would rather we prepare well or even avoid the possibility.

Max Tegmark, author of *Life 3.0* (2018), uses the example that we only built the fire extinguisher after the discovery of fire. This is quite general; however, if we take the view that Stuart Russell's idea in *Human Compatible* (2019) of AI working out what is best for humans, we will be setting off in the right direction – with Max Tegmark's AI fire extinguisher in hand just in case. Moving forward, as long as we aim to go in the right direction with our safety nets in place, we can harness the benefits of our objective AI or ML to help us deal with the complex world.

6.5 SUMMARY

This chapter explored the future of AI and its impact on careers, highlighting how AI and ML will evolve from the present to the distant future. The chapter discussed the emergence of new roles and career paths as AI continues to develop, along with human-only, machine-only and human-plus-machine systems. Human-only systems will focus on areas such as leadership, creativity and empathy, while machines excel at repetitive, predictable tasks and automation.

The chapter presented opportunities in roles such as data scientists, AI research scientists, computer vision engineers and AI ethics specialists, showcasing the expanding scope of AI-related careers. It also outlined the integration of AI into various industries, including healthcare, finance, education and entertainment, where AI is driving innovation and efficiency.

Looking ahead, the chapter considered future trends in AI, such as advancements in AGI, human–AI collaboration, AI ethics and quantum computing. It also delved into potential challenges, including the ethical implications of AI, loss of control and risks associated with autonomous systems.

The chapter concluded with a discussion on consciousness and AI, examining the complexities of developing machine consciousness and the potential ethical concerns that accompany such advancements. While AI holds immense promise, its future depends on careful governance, responsible development and a focus on enhancing human capabilities.

REFERENCES

Bommasani, R., Hudson, D.A., Adeli, E., et al. (2021) 'On the opportunities and risks of foundation models'. California: Stanford University. https://crfm.stanford.edu/report.html.

Bronshteĭn, I.N., Semendiaev, K.A., Musiol, G. and Mühlig, H. (2015) *Handbook of Mathematics*. Heidelberg and New York: Springer.

Brooks, F.P. (1986) 'No silver bullet: essence and accident in software engineering'. *Proc. IFIP Tenth World Computing Conference*, 1069–1076.

Cabinet Office, Government of Japan (2022) 'Social principles of human-centric AI'. Council for Social Principles. www8.cao.go.jp/cstp/stmain/aisocialprinciples.pdf.

Chalmers, D.J. (1996) *The Conscious Mind: In Search of a Fundamental Theory*. New York: Oxford University Press.

Coole, D. (2017) 'Agency: political science'. Britannica. https://britannica.com/topic/agency-political-theory.

Crick, F. (1994) *The Astonishing Hypothesis: The Scientific Search for the Soul*. Hoboken, NJ: Prentice Hall.

Cynefin (2023) 'Naturalising sense-making'. Cynfin.io. https://cynefin.io/wiki/Main_Page.

Daugherty, P.R. and Wilson, H.J. (2018) *Human + Machine: Reimagining Work in the Age of AI*. Brighton, MA: Harvard Business Review Press.

Dreyfus, H. (1972) *What Computers Can't Do*. New York: MIT Press.

Duignan, B. (2010) 'Venn diagram: logic and mathematics'. Britannica. https://britannica.com/topic/Venn-diagram.

European Commission (2018) 'GDPR'. European Commission. https://commission.europa.eu/law/law-topic/data-protection/data-protection-eu_en.

European Commission (2024) 'Ethics guidelines for trustworthy AI'. European Commission. https://digital-strategy.ec.europa.eu/en/library/ethics-guidelines-trustworthy-ai.

European Parliament (2023) 'The AI Act'. Topics: European Parliament. https://europarl.europa.eu/topics/en/article/20230601STO93804/eu-ai-act-first-regulation-on-artificial-intelligence.

Floridi, L. and Cowls, J. (2019) 'A unified framework of five principles for AI in society'. *HDSR*, 1 July. https://hdsr.mitpress.mit.edu/pub/l0jsh9d1/release/8.

Future of Life (2023) 'Pause giant AI experiments: an open letter'. https://futureoflife.org/open-letter/pause-giant-ai-experiments/.

Géron, A. (2017) *Hands-On Machine Learning with Scikit-Learn and TensorFlow: Concepts, Tools, and Techniques to Build Intelligent Systems*. Sebastopol, CA: O'Reilly Media.

Government of Canada (2024) 'Responsible use of artificial intelligence in government'. Canada.ca. https://canada.ca/en/government/system/digital-government/digital-government-innovations/responsible-use-ai.html.

Graham, R.L., Knuth, D.E. and Patashnik, O. (1994) *Concrete Mathematics*, 2nd edn. Boston, MA: Addison Wesley.

Gregersen, E. (2024) 'Big data'. Britannica. https://britannica.com/technology/big-data.

ISO/IEC (2018) '31000:2018 Risk management – guidelines'. iso.org. https://iso.org/standard/65694.html.

ISO/IEC (2022a) '22989:2022 Information technology – artificial intelligence – artificial intelligence concepts and terminology'. iso.org. https://iso.org/standard/74296.html.

ISO/IEC (2022b) '23053:2022 Framework for Artificial Intelligence (AI) Systems Using Machine Learning (ML)' iso.org. https://iso.org/standard/74438.html.

ISO/IEC (2023a) '42001:2023 Information technology – artificial intelligence – management system'. iso.org. https://iso.org/standard/81230.html.

ISO/IEC (2023b) '23894:2023 Information technology – artificial intelligence – guidance on risk management'. iso.org. https://iso.org/standard/77304.html.

Knuth, D.E. (2011) *The Art of Computer Programming*, vols 1–4a, revised edn. Boston, MA: Addison Wesley.

Kurzweil, R. (2012) *How to Create a Mind: The Secret of Human Thought Revealed*. New York: Viking Penguin.

Kurzweil, R. (2016) *The Singularity is Near: When Humans Transcend Biology*. London: Duckworth Overlook.

Lee Cooke, R. (2020) 'Andrey Nikolayevich Kolmogorov: Russian mathematician'. Britannica. https://britannica.com/biography/Andrey-Nikolayevich-Kolmogorov.

Lighthill, Sir J. (1973) 'Artificial intelligence: a general survey'. In: *Artificial Intelligence: A Paper Symposium*. London: Science Research Council.

Lu, D. (2019) 'Creating an AI can be five times worse for the planet than a car'. *New Scientist*, June. https://newscientist.com/article/2205779-creating-an-ai-can-be-five-times-worse-for-the-planet-than-a-car/.

McCarthy, J. (2000) 'Review of *What Computers Still Can't Do* by Hubert Dreyfus'. *Artificial Intelligence*, November. http://jmc.stanford.edu/artificial-intelligence/reviews/dreyfus.pdf.

McCulloch, W.S. and Pitts, W. (1943) 'A logical calculus of the ideas immanent in nervous activity'. *Bull. Math. Biophys.*, 5 (4), 115–133.

Minsky, M.L. and Papert, S.A. (1988) *Perceptrons: An Introduction to Computational Geometry*, ext. edn. Cambridge, MA: MIT Press.

Mitchell, T.M. (2018) *Machine Learning*, int. edn. New York: McGraw Hill Education.

Murphy, R. (2019) *Introduction to AI Robotics*. Cambridge, MA: MIT Press.

NIST (2023) 'Artificial Intelligence Risk Management Framework (AI RMF 1.0)'. US National Institute of Standards and Technology. https://nvlpubs.nist.gov/nistpubs/ai/NIST.AI.100-1.pdf.

NIST (2024) 'NIST Risk Management Framework: RMF'. US National Institute of Standards and Technology. https://csrc.nist.gov/Projects/risk-management/about-rmf.

OpenAI (2019) 'Solving Rubik's Cube with a robot hand'. OpenAI. https://openai.com/blog/solving-rubiks-cube/.

Penrose, R., Hameroff, S.R., Kak, S. and Tao, L. (eds) (2011) *Consciousness and the Universe: Quantum Physics, Evolution, Brain & Mind*. Cambridge, MA: Cosmology Science Publishers.

Pitts, W. and McCulloch, W.S. (1947) 'How we know universals the perception of auditory and visual forms'. *Bull. Math. Biophys.*, 9 (3), 127–147.

Python (2019) 'Homepage'. Python.org. https://python.org.

Rao, A.S. and Verweij, G. (eds) (2017) 'Sizing the prize: what's the real value of AI for your business and how can you capitalise?'. PwC Global. https://pwc.com/gx/en/news-room/docs/report-pwc-ai-analysis-sizing-the-prize.pdf.

Rosenblatt, F. (1962) *Principles of Neurodynamics: Perceptrons and the Theory of Brain Mechanisms*. Washington, DC: Spartan Books.

Rothman, D. (2024) *Transformers for Natural Language Processing and Computer Vision: Explore Generative AI and Large Language Models with Hugging Face, ChatGPT, GPT-4V, and DALL-E 3*, 3rd edn. Birmingham: Packt Publishing.

Routledge, R. (2005) 'Bayes's theorem: probability'. Brittanica. https://britannica.com/topic/Bayess-theorem.

Rumelhart, D.E. and McClelland, J.L. (1986) *Parallel Distributed Processing: Explorations in the Microstructure of Cognition*. Cambridge, MA: MIT Press.

Russell, S. (2019) *Human Compatible: Artificial Intelligence and the Problem of Control*. New York: Viking.

Russell, S. and Norvig, P. (2016) *Artificial Intelligence: A Modern Approach*, 4th edn. Harlow: Pearson Education Ltd.

Scikit-learn.org (2024) 'Scikit-learn'. https://scikit-learn.org/stable/.

Searle, J.R. (2002) *Consciousness and Language*. Cambridge: Cambridge University Press.

Seth, A. (2021) *Being You*. London: Faber & Faber Ltd.

Sternburg, R.J. (2017) 'Human intelligence: psychology'. Britannica. https://britannica.com/science/human-intelligence-psychology.

Strang, G. (2016) *Introduction to Linear Algebra*. Course available via MIT OpenCourseware. Wellesey-Cambridge Press. https://ocw.mit.edu/courses/mathematics/18-06sc-linear-algebra-fall-2011/syllabus/.

Strang, G. (2019) *Linear Algebra and Learning from Data*. Wellesley, MA: Wellesey-Cambridge Press.

Tegmark, M. (2018) *Life 3.0: Being Human in the Age of Artificial Intelligence*. London: Penguin Books.

TensorFlow (2019) 'Homepage'. https://tensorflow.org.

Teukolsky, S.A., Vetterling, W.T., Press, W.H. and Flannery, B.P. (1993) *Numerical Recipes in Fortran*, 2nd edn. Cambridge: Cambridge University Press.

The Alan Turing Institute (2024) 'AI Standards Hub'. https://aistandardshub.org.

The Editors of Encyclopaedia Britannica (2024a) 'Dorothy Vaughan: American mathematician'. Britannica. https://britannica.com/biography/Dorothy-Vaughan.

The Editors of Encyclopaedia Britannica (2024b) 'Probability density function: mathematics'. Britannica. https://britannica.com/science/density-function.

The White House (2022) 'Blueprint for an AI Bill of Rights'. The White House. https://whitehouse.gov/ostp/ai-bill-of-rights/.

Turing, A.M. (1950) 'Computing machinery and intelligence'. *Mind*, 49 (236), 433–460.

UK Government (2018) 'Data Protection Act'. GOV.UK. https://gov.uk/data-protection.

United Nations (2015) 'The sustainable development agenda'. UN.org. https://un.org/sustainabledevelopment/development-agenda/.

Vincent, J. (2019) 'Former Go champion beaten by DeepMind retires after declaring AI invincible'. *The Verge*, 27 November. https://theverge.com/2019/11/27/20985260/ai-go-alphago-lee-se-dol-retired-deepmind-defeat.

Warnock, Lady M. (2006) *An Intelligent Person's Guide to Ethics*. London: Gerald Duckworth & Co. Ltd.

Wikipedia Contributors (2024a) 'Ada Lovelace'. Wikipedia. https://en.wikipedia.org/wiki/Ada_Lovelace.

Wikipedia Contributors (2024b) 'René Descartes'. Wikipedia. https://en.wikipedia.org/wiki/René_Descartes.

Wikipedia Contributors (2024c) 'Leonardo da Vinci'. Wikipedia. https://en.wikipedia.org/wiki/Leonardo_da_Vinci.

Wikipedia Contributors (2024d) 'Neil Armstrong'. Wikipedia. https://en.wikipedia.org/wiki/Neil_Armstrong.

Wikipedia Contributors (2024e) 'Karen Spärck Jones'. Wikipedia. https://en.wikipedia.org/wiki/Karen_Spärck_Jones.

Wikipedia Contributors (2024f) 'Tu Youyou'. Wikipedia. https://en.wikipedia.org/wiki/Tu_Youyou.

Wikipedia Contributors (2024g) 'Scientific method'. Wikipedia. https://en.wikipedia.org/wiki/Scientific_method.

Wikipedia Contributors (2024h) 'Roger Bannister'. Wikipedia. https://en.wikipedia.org/wiki/Roger_Bannister.

Wikipedia Contributors (2024i) 'Digital twin'. Wikipedia. https://en.wikipedia.org/wiki/Digital_twin.

Wikipedia Contributors (2024j) 'Asilomar conference on beneficial AI'. Wikipedia. https://en.wikipedia.org/wiki/Asilomar_Conference_on_Beneficial_AI.

Wikipedia Contributors (2024k) 'Laws of robotics'. Wikipedia. https://en.wikipedia.org/wiki/Laws_of_robotics.

Wikipedia Contributors (2024l) 'Agent-based model'. Wikipedia. https://en.wikipedia.org/wiki/Agent-based_model.

Wikipedia Contributors (2024m) 'Subsumption architecture'. Wikipedia. https://en.wikipedia.org/wiki/Subsumption_architecture.

Wikipedia Contributors (2024n) 'Gaussian elimination'. Wikipedia. https://en.wikipedia.org/wiki/Gaussian_elimination.

Wikipedia Contributors (2024o) 'Statistical learning theory'. Wikipedia. https://en.wikipedia.org/wiki/Statistical_learning_theory.

Wikipedia Contributors (2024p) 'Agile management'. Wikipedia. https://en.wikipedia.org/wiki/Agile_management.

Wikipedia Contributors (2024q) 'Waterfall model'. Wikipedia. https://en.wikipedia.org/wiki/Waterfall_model.

FURTHER READING

Géron, A. (2019) *Hands-On Machine Learning with Scikit-Learn, Keras and TensorFlow: Concepts, Tools, and Techniques to Build Intelligent Systems*, 2nd edn. Sebastopol, CA: O'Reilly Media.

Schwab, K. (2016) *The Fourth Industrial Revolution*. London: Penguin Random House.

Theobald, O. (2017) *Machine Learning for Absolute Beginners: A Plain English Introduction*. London: Scatterplot Press.

USEFUL WEBSITES

ARTIFICIAL INTELLIGENCE

Open AI

> https://openai.com/

European Parliament Committees

> https://europarl.europa.eu/committees/en/indexsearch?scope=CURRENT&term=
> 9&query=artificial+intelligence&scope=ALL&ordering=RELEVANCE

Cyber Defence

> https://eda.europa.eu/webzine/issue14/cover-story/artificial-intelligence-(ai)-
> enabled-cyber-defence

UK Office for Science, Innovation and Technology

> https://gov.uk/government/publications/national-ai-strategy

ETHICS

The Turing Institute

> https://turing.ac.uk/research/data-ethics

European Union

> https://ec.europa.eu/digital-single-market/en/news/ethics-guidelines-trustworthy-ai

Future of Life Institute

>   https://futureoflife.org/

MACHINE LEARNING

The Royal Society

>   https://royalsociety.org/topics-policy/projects/machine-learning/videos-and-background-information/

Google for Education: Python

>   https://developers.google.com/edu/python

SUSTAINABILITY

United Nations

>   https://un.org/sustainabledevelopment/sustainable-development-goals/

>   https://un.org/en/development/desa/policy/cdp/cdp_publications/2012cdppolicynote.pdf

Smart Cities

>   https://smartcitiesworld.net/

International Organization for Standardization

>   https://iso.org/home.html

GLOSSARY

Artificial intelligence (AI): Artificial intelligence refers to systems designed by humans that, given a complex goal, act in the physical or digital world by perceiving their environment, interpreting the collected structured or unstructured data, reasoning on the knowledge derived from these data and deciding the best action(s) to take (according to predefined parameters) to achieve the given goal. AI systems can also be designed to learn to adapt their behaviour by analysing how the environment is affected by their previous actions.

As a scientific discipline, AI includes several approaches and techniques, such as machine learning (of which deep learning and reinforcement learning are specific examples), machine reasoning (which includes planning, scheduling, knowledge representation and reasoning, search and optimisation) and robotics (which includes control, perception, sensors and actuators, as well as the integration of all other techniques into cyber-physical systems).

Assistive robot: A robot designed to provide physical or cognitive assistance to a person.

Autonomy: The ability of a machine to make its own decisions.

Axon terminals: Axon terminals are terminations of the telodendria (branches) of an axon.

Backpropagation: A method used in artificial neural networks to calculate a gradient required in the calculation of the weights to be used in the network.

Bias: The deviation of the expected value of a statistical estimate from the quantity it estimates. It can also be considered to be a prejudice for or against something or somebody that may result in unfair decisions. It is known that humans are biased in their decision-making. Since AI systems are designed by humans, it is possible that humans inject their bias into the systems, even in an unintended way. Many current AI systems are based on machine learning data-driven techniques. Therefore, a predominant way to inject bias can be in the collection and selection of training data. If the training data are not inclusive and balanced enough, the system could learn to make unfair decisions. At the same time, AI can help humans to identify their biases, and assist them in making less biased decisions.

Big data: Data sets that are so big and complex that traditional data-processing application software is inadequate to deal with them.

Chatbot: An artificial intelligence program that conducts a conversation via auditory or textual methods.

Classification: The problem of identifying to which of a set of classes a new observation belongs.

Clustering: Groups a set of objects in such a way that objects in the same group are more similar to each other than to those in other groups.

Combinatorial explosion: The rapid growth of the complexity of a problem due to the combinations of the problem's input parameters.

Data analysis: The process of inspecting, cleaning, transforming and modelling data with the goal of discovering useful information, drawing conclusions and supporting decision-making. Data analysis involves applying various techniques, such as statistical analysis, data visualisation and machine learning, to identify patterns, trends, relationships and insights from raw data.

Data cleaning: Detects and corrects (or removes) corrupt or inaccurate records from a record set, table or database and refers to identifying incomplete, incorrect, inaccurate or irrelevant parts of the data and then replacing, modifying or deleting the dirty or coarse data.

Data mining: The process of discovering patterns in large data sets.

Data science: Uses scientific methods, processes, algorithms and systems to understand data.

Data scrubbing: See data cleaning.

Decision trees: A decision support tool that uses a tree-like graph or model of decisions and their possible consequences.

Deep learning: A class of algorithms that use a cascade of multiple layers for feature extraction and transformation. Each successive layer uses the output from the previous layer as input.

Dendrites: Branched extensions of a nerve cell that propagate the electrochemical stimulation.

Emotional intelligence or emotional quotient (EQ): The understanding of our emotions and the emotions of others.

Ensemble: Methods that use multiple learning algorithms to obtain better predictive performance than could be obtained from any of the constituent learning algorithms alone.

Ethical purpose: Used to indicate the development, deployment and use of AI, which ensures compliance with fundamental rights and applicable regulation as well as respecting core principles and values. This is one of the two core elements to achieve trustworthy AI.

143

Expert systems: A computer system that emulates the decision-making ability of a human expert.

Feedforward neural network: An artificial neural network wherein connections between the nodes do not form a cycle.

Functionality: The tasks that a computer software program is able to do.

Hardware: The physical parts or components of a computer.

Heuristic: A strategy derived from previous experiences with similar problems.

Human-centred AI (HCAI): An approach to AI development that puts humans at the forefront by designing systems that are user-friendly, transparent, explainable and ethically responsible. It aims to create AI that collaborates with humans in ways that are understandable, controllable and beneficial, ensuring that AI solutions are designed to improve human experiences, safety and wellbeing.

Human-centric AI: AI that emphasises enhancing and empowering human abilities by ensuring that AI systems work alongside humans, augmenting their capabilities and supporting decision-making processes. It focuses on aligning AI with human values, ethical considerations and societal needs. Human-centric AI is more about collaboration and augmentation, whereas human-centred AI emphasises usability, transparency and ethical design.

Hyper-parameter: A parameter whose value is set before the learning process begins.

Inductive reasoning: Makes broad generalisations from specific observations.

Internet of Things (IoT): The network of physical devices, vehicles, home appliances and other items embedded with electronics, software, sensors, actuators and connectivity that enable these things to connect and exchange data.

K-means: A clustering algorithm that partitions observations into k clusters in which each observation belongs to the cluster with the nearest mean, serving as a prototype of the cluster.

K-nearest neighbours: The simplest clustering algorithm used to classify new data points based on the relationship to nearby data points.

Layers: Neural networks are organised into layers and a layer is a set of interconnected nodes.

Linear algebra: The branch of mathematics concerning linear equations and functions and their representations through matrices and vector spaces.

Logistic regression: Used in binary classification to predict two discrete classes.

Machine learning (ML): A subset of artificial intelligence in the field of computer science that gives computers the ability to learn from data.

Narrow AI or weak AI: Narrow artificial intelligence, also known as weak artificial intelligence, is AI focused on one narrow task. It is the contrast of strong AI.

Natural language processing (NLP): An area of artificial intelligence that focuses on the interaction between computers and humans through natural language. NLP enables machines to understand, interpret and generate human language, allowing them to perform tasks such as language translation, sentiment analysis, speech recognition and text generation. It combines computational linguistics, machine learning and deep learning techniques to process and analyse large volumes of natural language data.

Nearest neighbour algorithm: One of the first algorithms used to determine a solution to the travelling salesman problem.

Neural network: A machine learning algorithm that is based on a mathematical model of the biological brain.

Nodes: Represent neurons (biological brain) and are interconnected to form a neural network.

Ontology: The philosophical study of the nature of being, becoming, existence or reality, as well as the basic categories of being and their relations.

Optical character recognition (OCR): The conversion of images of typed, handwritten or printed text into machine-encoded text.

Over-fitting or over-training: A machine learning model that is too complex, has high variance and low bias. It is the opposite of under-fitting or under-training.

Probability: The measure of the likelihood that an event will occur.

Python: A programming language popular in machine learning.

Random decision forests: An ensemble learning method for classification, regression and other tasks.

Random forests: An ensemble learning method for classification, regression and other tasks that operate by constructing a multitude of decision trees at training time.

Regression analysis: In machine learning, regression analysis is a simple supervised learning technique used to find a trendline to describe the data.

Reinforcement machine learning: Uses software agents that take actions in an environment in order to maximise some notion of cumulative reward.

Robotic process automation (RPA): A business process automation technology based on the notion of software robots or artificial intelligence workers.

Robotics: Deals with the design, construction, operation and use of robots, as well as computer systems for their control, sensory feedback and information processing.

Scripting: Programs written for a special run-time environment that automate the execution of tasks that could alternatively be executed one-by-one by a human operator.

Search: The use of machine learning in search problems, e.g. shortest path.

Semi-supervised machine learning: Machine learning that uses labelled and unlabelled data for training.

Sigmoid function: A mathematical function having a characteristic S-shaped curve or sigmoid curve.

Software: A generic term that refers to a collection of data and computer instructions that tell the computer how to work.

Software robot: Replaces a function that a human would otherwise do.

Strong AI or artificial general intelligence: Strong AI's goal is the development of artificial intelligence to the point where the machine's intellectual capability is functionally equal to a human.

Supervised machine learning: The task of learning a function that maps an input to an output based on example input–output pairs.

Support vector machine: Constructs a hyperplane or set of hyperplanes in a high- or infinite-dimensional space, which can be used for classification, regression or other tasks, such as an outlier's detection.

Swarm intelligence: The collective behaviour of decentralised, self-organised systems, natural or artificial.

System: A regularly interacting or interdependent group of items forming a unified whole.

The fourth industrial revolution: Builds on the digital revolution, representing new ways in which technology becomes embedded within societies and even the human body.

Trustworthy AI: Has two components: (1) its development, deployment and use should comply with fundamental rights and applicable regulation as well as respecting core principles and values, ensuring 'ethical purpose'; (2) it should be technically robust and reliable.

Under-fitting: When the machine learning model has low variance and high bias. It is the opposite of over-fitting or over-training.

Universal design: Refers to broad-spectrum ideas meant to produce buildings, products and environments that are inherently accessible to older people, people without disabilities and people with disabilities (close relation to inclusive design).

Unsupervised machine learning: Infers a function that describes the structure of unlabelled data.

Validation data: A set of data used to test the output of a machine learning model that is not used to train the model.

Variance: The expectation of the squared deviation of a random variable from its mean.

Visualisation: Any technique for creating images, diagrams or animations to communicate a message.

Weights: A weight function is a mathematical device used when performing a sum, integral or average to give some elements more 'weight' or influence on the result than other elements in the same set.

INDEX

Page numbers in italics refer to figures or tables.

strong AI 7, 8, 13, 18, 10

structured state 38, *39*

subjective experience 2, 8, 131, 132

subjectivity 2

super-intelligent AI 7, 13, 130

supervised learning 54, *55*, 70, 71, 72, 73, 74

support vector machines (SVM) 7, 54, 71, 73, 84, 103

sustainability 6, 16, *17*, 22, 89, 128, 129

SWOT analysis 29, *29*

system of systems 6

Tegmark, Professor Max 8, 14, 101, 132

Tensorflow 61

theory of mind 13

tokenisation 83

transformer models 12, 82

transparency 19–22, 24–8, 31–2, 88–9, 93, 96–7, 106–9, 111

Turing, Alan 9

UN sustainability goals 19, 23

unacceptable risk AI systems 25

under-fitting 80, *81*, *82*

Unimate 41

universal design 15

unsupervised learning 7, 12, 54, 55, *56*, 70, 71, 103

user empowerment 31, 109

user engagement 34, 123

user experience (UX) 7, 31, 33, 106, 109, 111, 112, 126

user-centred design 107, 108

utility-based reflex agent 40

variance error 80

variational autoencoder (VAE) 7, 82

Vaughan, Dorothy 48, 49

vector calculus 47–9, 61, 79

virtual assistants 33, 100, 105, 113, 115, 117, 119, 125, 126, 128

virtual reality (VR) 61, 63, 105, 120, 123, 129

Waterfall project management 93–5

WCAG (web content accessibility guidelines) 24, 25

What Computers Can't Do 132

Wilson, H James 85, 101

www.ingramcontent.com/pod-product-compliance
Lightning Source LLC
Chambersburg PA
CBHW041008050326
40690CB00031B/5300